NORTH WEST CANALS

Merseyside, Weaver & Chester

THROUGH TIME

Ray Shill

AMBERLEY PUBLISHING

First published 2013

Amberley Publishing
The Hill, Stroud, Gloucestershire, GL5 4EP
www.amberley-books.com

ISBN 978 1 4456 1891 3 (print)
ISBN 978 1 4456 1903 3 (ebook)

British Library Cataloguing in Publication Data.
A catalogue record for this book is available from the British Library.

Typesetting by Amberley Publishing.
Printed in Great Britain.

Contents

Introduction

This is the third volume in a series that investigates waterways in Britain, focusing particularly on canal and river navigation infrastructure. While much has disappeared, important features remain as a testament to the engineering and construction skills of those that made these works possible.

Photographs and other images are primarily drawn from the author's collection or that of the Railway and Canal Historical Society (RCHS), whose members, over time, have contributed an important group of images to the society collection. Several photographers are listed as contributors to this collection. One of the most prolific for the waterways of the North West was P. Norton, who recorded many features that have disappeared, particularly in the Warrington district.

The North West might be said to have encouraged a pioneering spirit in canal transport. It was here that some of the earliest canals were made. There was a mixture of river navigations and canals that served the industries in this part, which ranged from mining to cotton spinning. River craft were often referred to as 'Mersey Flats' or simply 'flats', even if their range extended beyond the Mersey to the Irwell and Dee, as well as along the canals, which were made to barge width dimensions.

Canal navigations linked with the rivers at Chester, Ellesmere Port, Runcorn, Liverpool and Warrington. They formed internal waterway routes across the Pennines into Wales, and southwards through the Potteries to the Midlands and London. The Port of Liverpool was at the centre of this trade where flats exchanged cargos with coastal and seagoing vessels. It was an age when transport was labour intensive and many forms and types of goods were transported by water. Raw materials came into Liverpool and finished goods left. It was seemingly an endless cycle built on industry. Most important was the cotton imports that passed to the mill towns such as Manchester along the Mersey & Irwell Navigation, or the china clay destined for the potteries that found its way towards Runcorn before passing on by narrowboat for the last leg of the trip. Outward cargoes onto the Mersey and Weaver included salt and pottery; inward trade included cotton, china clay, ironstone and pig iron.

On Deeside, quays and wharves were once regular calling points for seagoing and coastal vessels that came down the Dee, the channel of which was improved in the eighteenth century. Flats from 1779 were able to join the Chester Canal, which was a barge waterway built to Nantwich. A nineteenth-century development was at Saltney where quays and a landing stage were built beside the Dee, and rail sidings put down for interchange of traffic between vessels and the Great Western Railway.

In this first photographic journey of North West waterways, Merseyside is investigated. A second book will cover Manchester and the Peak District.

Chapter One

River Navigations

Bromborough Pool, Dee, Ditton Brook, Douglas, Mersey, Ribble, Sankey Brook and Weaver

Rivers provided the first means of transporting goods inland, and several had their navigation extended through the provision of locks and weirs – the Dee, before it silted up, enabled navigation to Chester, while the Ribble was navigable as far as Penwortham Bridge, Preston. The Mersey was an important navigation and tidal limits reached Warrington, but required locks beyond there for craft to travel up towards Manchester. This combined Mersey & Irwell Navigation reached Salford and the rocky cliff on the Manchester side of the Irwell had parts cut away for wharves.

There were also inlets and wharves placed around the Mersey estuary, and some like the Ditton Brook could pass craft for a limited distance. The Ditton Brook Ironworks had a wharf on the brook to handle its cargoes. The Sankey Brook was also navigable from the Mersey as far as the Turnpike bridge, yet most of these inlets had restricted navigation that was tide dependent. Craft were frequently grounded at neap tides. On the Cheshire bank, Bromborough Pool was navigable for craft that served the pier and railway from Storeton Quarries. The Dee was straightened and improved during the eighteenth century. Vessels moored by a group of wharves and quays along the north/east bank, which was close to the city and the modern racecourse.

From the Ribble, the River Douglas was made navigable, with locks towards Wigan, serving several mines including those around Crooke. While the act for making the Douglas navigable was granted in 1720, limited finance put a stop to construction after only one lock, the tide lock at Croston Finney, and a section of river had been straightened. Work restarted from 1737 under new proprietors, and the task of making the river navigable to Wigan by what is believed to be thirteen locks was finally finished in 1743, although trade had been on the river for some two years already. Construction work included deepening the riverbed, shortening the river course at bends, and the provision of short lock cuts. The locks were made of stone and brick.

The Weaver also had locks, at first made from wood, then replaced by masonry structures, which enabled craft to travel as far as Winsford. This navigation was improved from time to time – lock improvement, additional locks, lock cuts, straightening, dredging and widening eventually produced a navigation that could pass large vessels. The number of locks finally increased to fourteen in total, but this number was reduced with the continuing improvement.

A key personage in engineering and/or planning many early navigations in this region was Thomas Steers. His lock schemes on the Douglas, Irwell and Mersey came to be adopted in one way or another, either with himself there surveying and engineering the works or others modifying his plans. In his day, Thomas Steers was recognised as an important engineer and surveyor, and the original docks at Liverpool were laid out by him. His achievements should be better recognised, as through him the foundations of the Merseyside waterways network were established.

Since Steer's time, these rivers have had their courses altered through improvement and drainage schemes. Looking at the modern river courses it is sometimes difficult to appreciate what was there, particularly the Mersey, where the Manchester Ship Canal construction enforced major diversions.

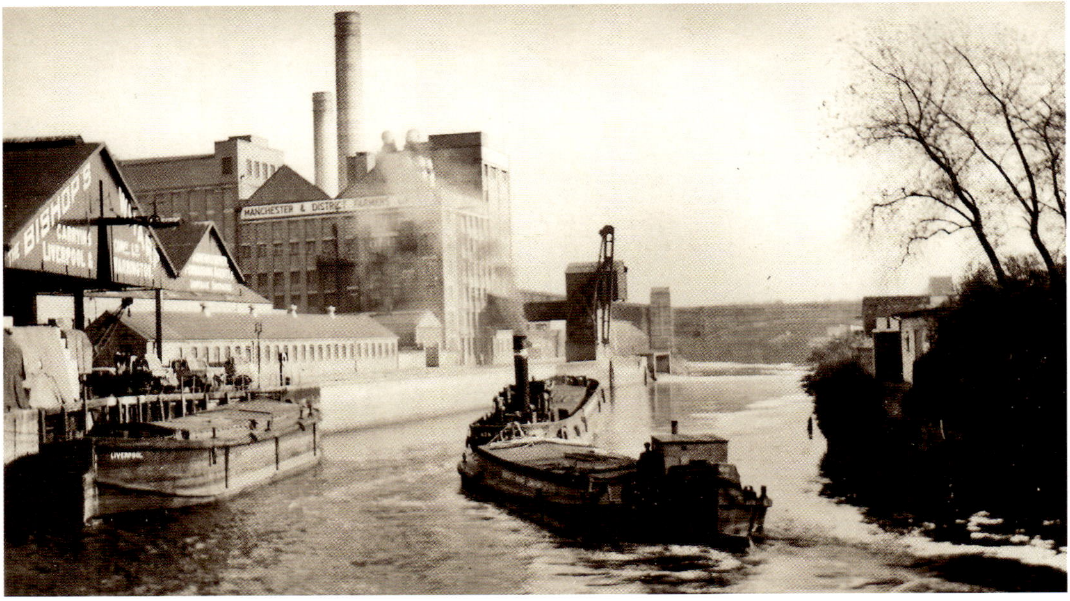

Bishops Wharf, River Mersey, Warrington
The Mersey from Bank Quay to Howley formed part of the Mersey & Irwell Navigation to Manchester; although, with the making of the Latchford to Runcorn Canal, it was relegated to being a branch off the main route between Runcorn and Manchester. (*RCHS Hugh Compton Collection 65186 & 677056*)

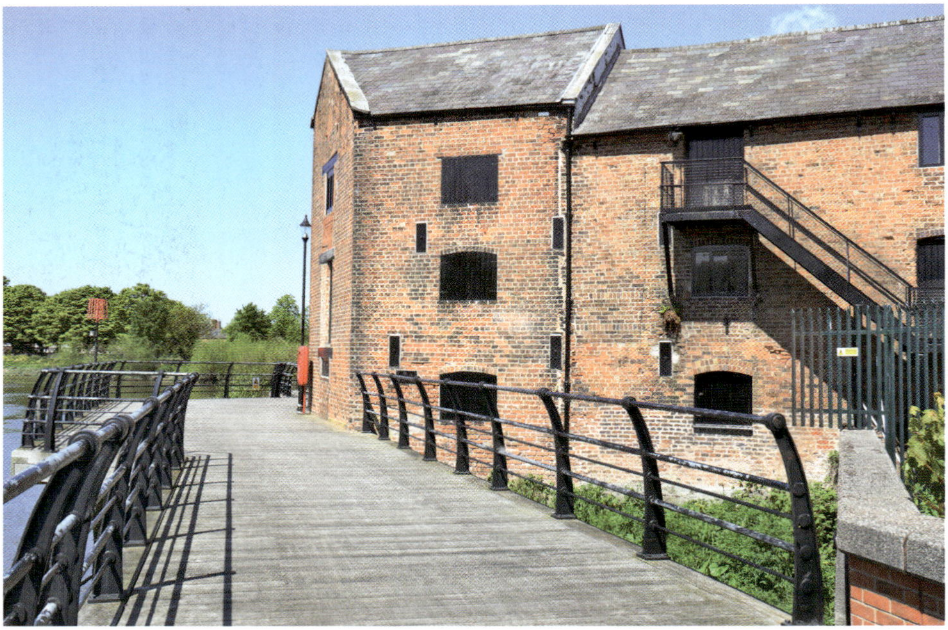

Former Warehouse, River Dee Wharf, Chester
New housing developments now surround this survivor. The long warehouse remains on Crane Wharf, River Dee, Chester, while the wharf space adjacent is a convenient spot for parking for the nearby racecourse. (*Ray Shill*)

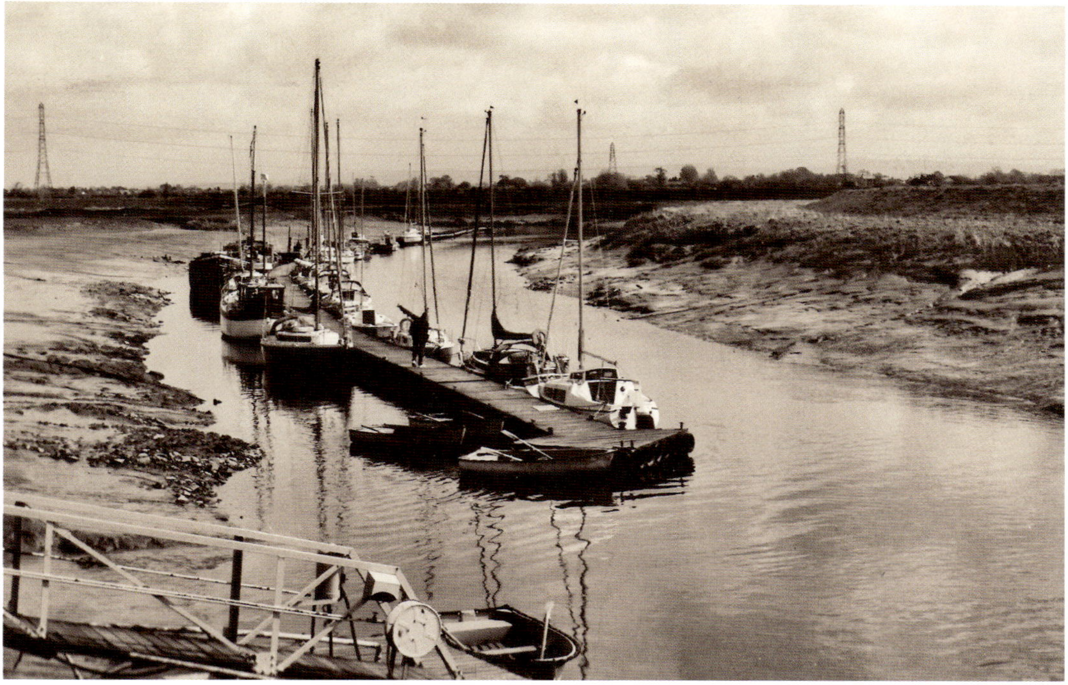

Hesketh Bank, River Douglas (*Above*) and Sollom (*Below*)
The section from the Ribble through Hesketh Bank to Tarleton remained navigable to the present time and now forms part of the Ribble Link to Preston. In previous times, the Douglas was navigable as far as mines near Wigan. A section of the Douglas between Tarleton and Sollom was canalised during 1760 and Sollom Lock remains. (*Ray Shill*)

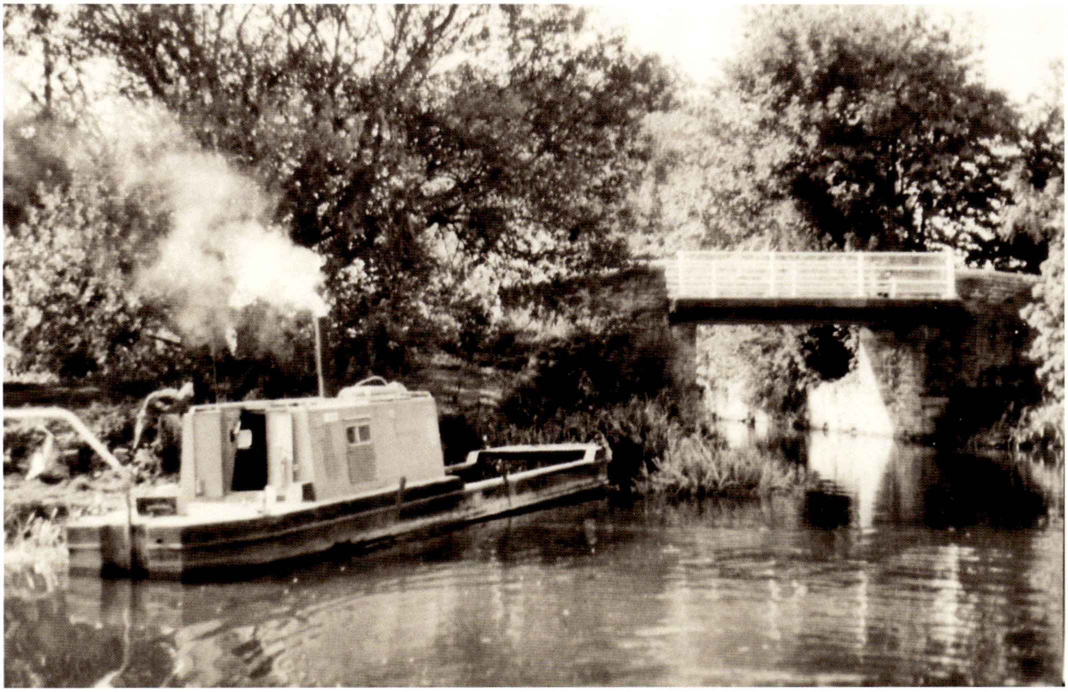

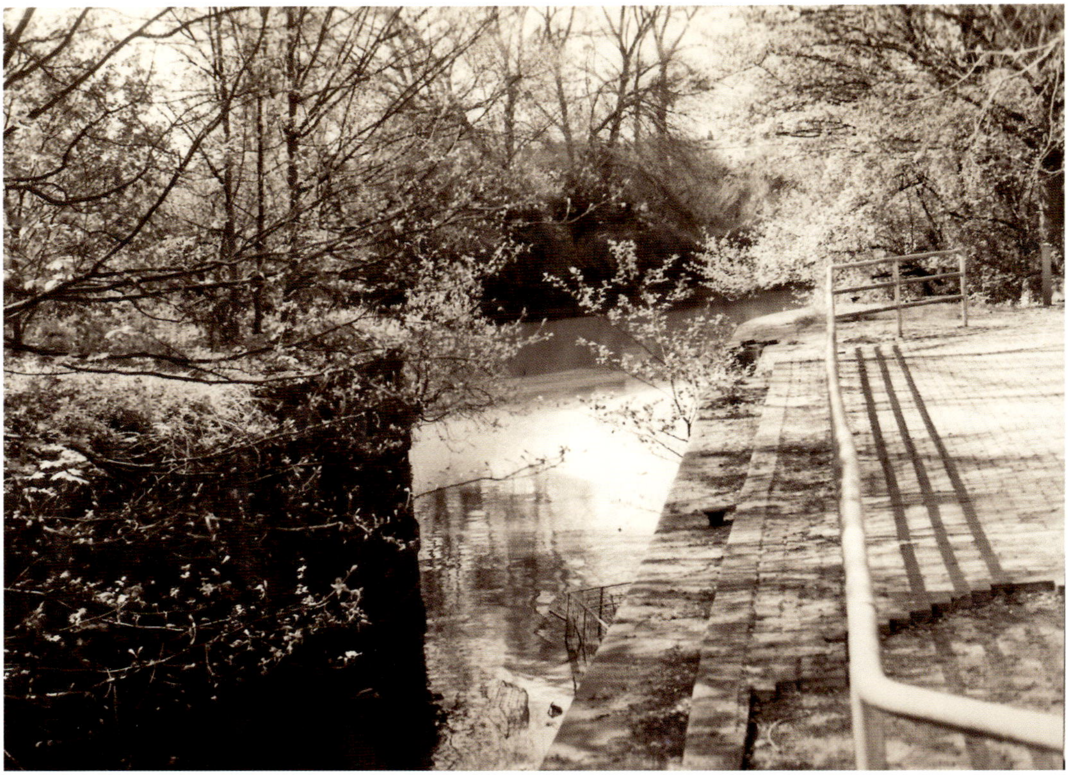

Howley (Victoria) Lock and Weir, Mersey & Irwell Navigation
Howley was the first lock on the Mersey & Irwell Navigation to Manchester, and was retained as a lock for craft serving Howley Quay or passing through to the Woolston Cut after the river was altered for the making of the Manchester Ship Canal. It was also located at the tidal limit for the Mersey. (*Above: Ray Shill. Below: RCHS Hugh Compton Collection 65192*)

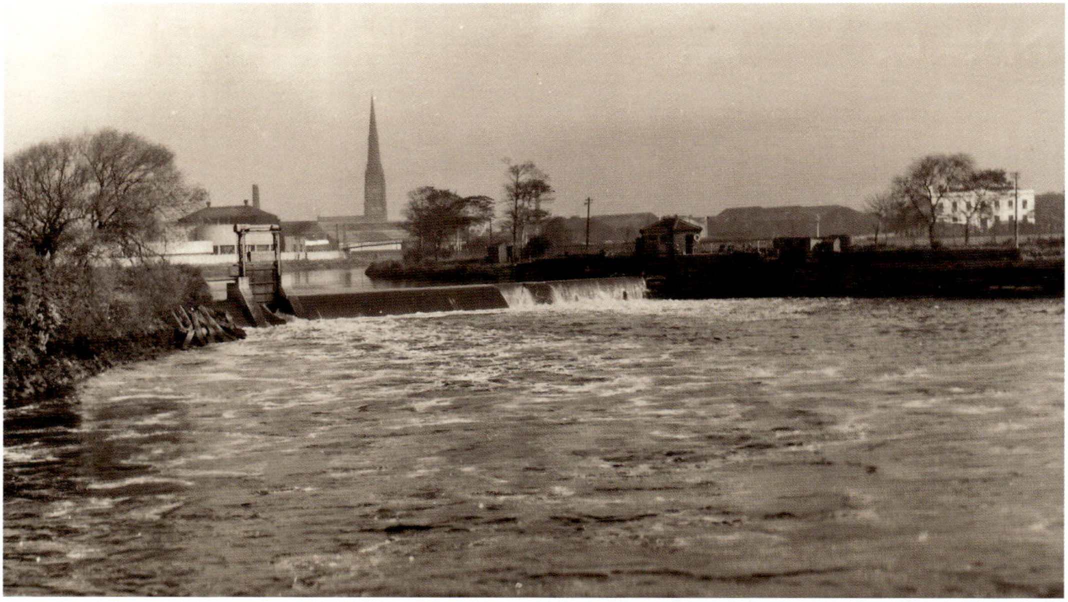

Chapter Two

Early Canals in the North West

Bridgewater, Chester, Sankey, Trent & Mersey

The age of the navigable canal in Britain may be said to have begun in Northern Ireland with the Newry Canal, but in mainland Britain that accolade can be awarded to the Sankey Brook Navigation, which joined the Mersey initially near Warrington and travelled northwards towards St Helens with several locks along the route. Thomas Steers (*c.* 1672–1750) was responsible for engineering the route of the Newry, while his assistant on the Newry Canal and Liverpool docks, Henry Berry (1720–1812), had the task of surveying the original navigation from Sankey Lock on the Sankey Brook to St Helens. The main aim of this canal was to transport coal from the mines at St Helens.

An act to make the Sankey Brook navigable received royal assent in 1755, but Henry Berry and his assistant on this project, William Taylor, chose a radical course of action. Instead of widening and deepening the narrow brook channel, a new line of canal was made parallel to the brook from Sankey Lock. The bulk of the work, complete with locks, as far as Parr and St Helens, was finished within two years. The streams that fed Sankey Brook were diverted into the canal and their lower reaches adapted as branches of the canal. Their water supply capacity was equivalent to a reservoir for the canal.

The outlet into Sankey Brook was hampered by the tide there, and navigation was restricted to the times when water levels were high. A canal extension act empowered the making of a canal from Sankey Lock to Fiddler's Ferry Lock on the Mersey where deeper water was available.

With the making of the Sankey Brook Navigation (otherwise known as the St Helens Canal, or Sankey Canal), a major step was made in the establishment of inland navigation across Britain. It was not long before another canal, the first section of the Bridgewater that ran from Castlefields to Worsley would be constructed. While the inspiration for the Bridgewater lay with the Duke of Bridgewater's observation of existing canals in France, the benefits of the local Sankey Navigation must have been known.

Like the Sankey, a prime reason for construction was the carriage of coal, which in this case was mined at Worsley where it came to the underground mines. This section was lockless but involved a crossing of the Irwell at Barton by an aqueduct. Water supply for the waterway included the mines and the River Medlock in Manchester, which flowed into Castlefield Basin. The Bridgewater was then extended to Runcorn and also formed a link with the Trent & Mersey Canal at Preston Brook.

This canal was level from Manchester to Runcorn, but was not completed until March 1776. At Runcorn, a section had opened earlier (1773) and included the locks to Mersey, where a descent of around 90 feet was achieved through the making of ten locks arranged as pairs of staircases and supplied with water stored in reservoirs near the top of the flight.

The branch to Preston Brook had important interchange warehouses, where canal craft from the narrowboat network of waterways might exchange cargoes with the flats for the journey across to the Port of Liverpool. East of these wharves was the boundary between the Bridgewater and Trent & Mersey. It was also the meeting point of the wide and narrow waterways, as the canal route soon passed into the curved and winding Preston Brook Tunnel. There were two more winding tunnels at Barnton and Saltisford before the canal came out onto the long open canal to Anderton. The route followed the Weaver Valley, but at a higher level to the Weaver. At Anderton the canal came close to the river and it was here that incline railways were subsequently built to trans-ship salt from the canal to the river. The canal at Runcorn descended through a flight of locks to a tidal basin, which gave access to the Mersey.

The Chester was a barge waterway that linked the Dee at Chester with the salt town of Nantwich. At the Dee the link was made into a long and wide tidal basin that had single gates across the entrance. It was a route with several locks, which included a staircase pair at Bunbury and, initially, a rise of five locks to Northgate Bridge from the Dee. The canal terminated at a basin at Nantwich, despite hopes that it might have continued to Middlewich. Expended revenue from tolls did not happen, and the undertaking had finance issues which restricted development.

A lesser known waterway was that built in 1768 for John Glynne, a Deeside landowner. It was only a mile long and brought coal down to the Dee at Saltney, although it did not link with the river. Coal was evidently trans-shipped into flats lying in an adjacent inlet.

Making the link between the Chester and the River Dee did not proceed smoothly:

Chester Chronicle, Thursday 25 April 1776

Chester Assizes

Thursday came on to be tried, at the above assizes, by a special jury, a cause wherein the River Dee Company were plaintiffs, and the Canal Company defendants. It was an issue directed by the High Court of Chancery, to try whether the plaintiffs had or had not complied with the terms of the Canal Act of Parliament respecting that part of the Canal, and Lock &c, which was made by them through their own lands. When, after some time spent in the hearing, a compromise was agreed to, upon the following terms: The Canal Company to pay to the River Dee Company, the expense of making the present flood-gate and works; to pay also a half-penny per ton for all goods, in goods wider than seven feet, passing to or from the River Dee, through the Dee Company's lands into, or out of, the Canal Lock, near the Water Tower, except stones, gravel, limestone, potters clay, slates and flint stones, which are free of tonnage. On these terms the Canal Company are to have a new pair of dock gates, and other works, erected at their own expence, so as to admit vessels of 27 feet in width into the intended bason on the Dee Company land, and vessels of 14 feet in width to navigate into and out of the Canal. Each party to pay their own costs.

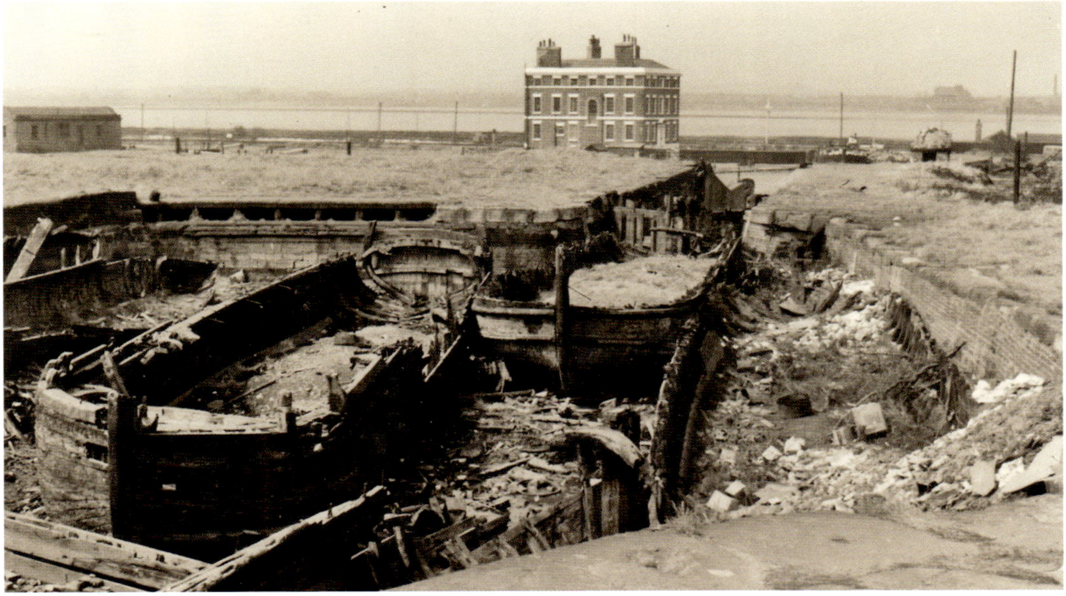

Old Runcorn Locks & Bridgewater House, Bridgewater Canal, 1957

The original lock flight was the first to be disused, becoming a repository for abandoned craft. In this view, Bridgewater House is seen by Manchester Ship Canal. Built in 1771, this house was initially a residence for the Duke of Bridgewater, but subsequently became the offices of the Bridgewater Trustees. In the lower image Bridgewater House is shown under restoration in 1998. (*Above: Railway & Canal Historical Society, Hugh Compton Collection, P. Norton 64294. Below: Ray Shill*)

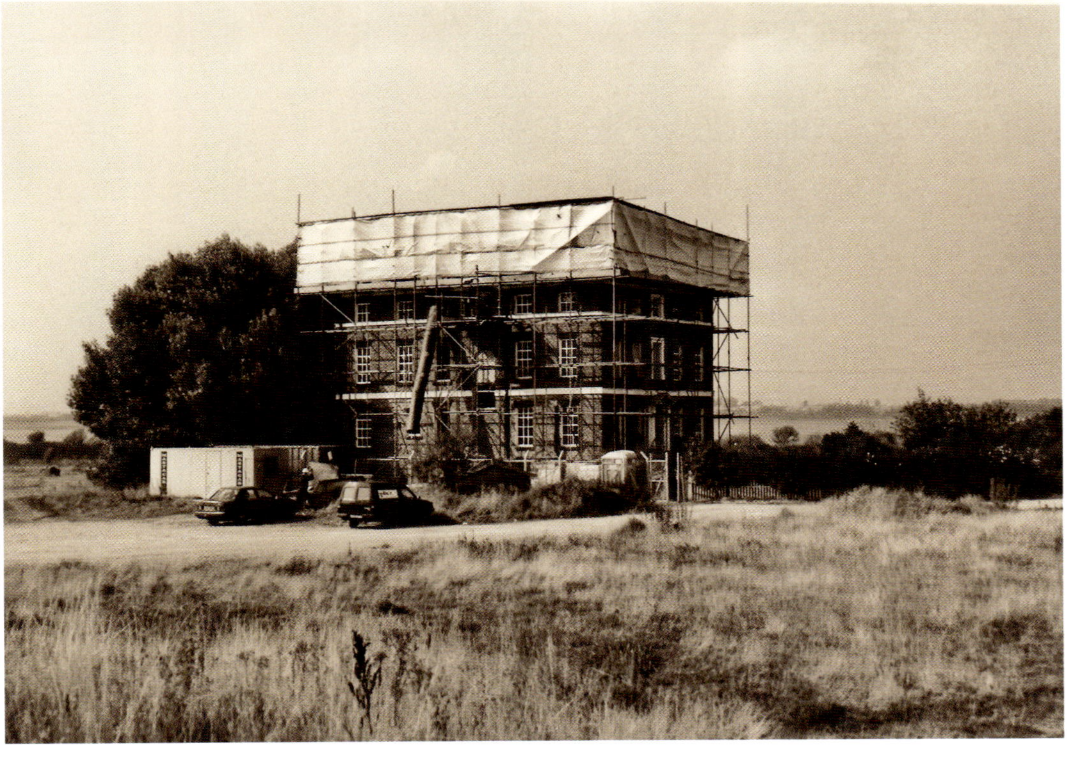

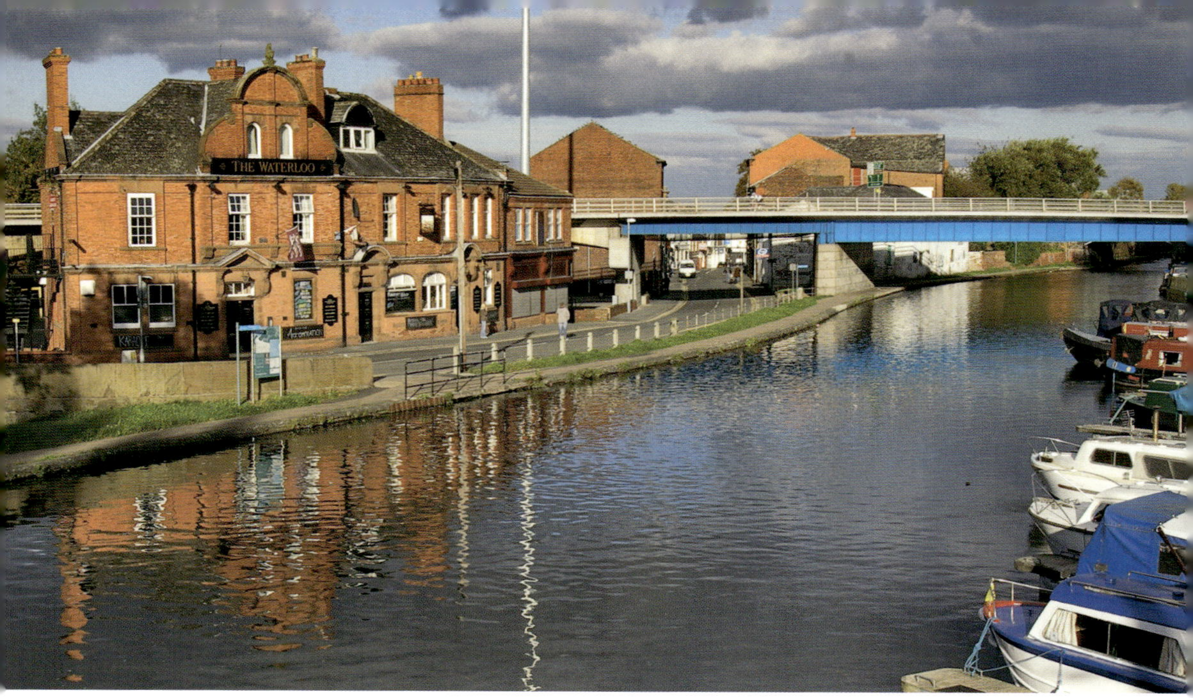

Runcorn Branch, Bridgewater Canal

The present terminus for the canal at Runcorn is placed east of the paired lock flights. It was the basin where flats and narrowboats waited their turn to descend to the docks at Runcorn. The top locks are partly buried now, as is the original lock flight down to the Mersey, although a restoration scheme has been proposed for craft to travel round to the Weston Canal and form a boating ring which would also involve the Anderton Lift. (*Ray Shill*)

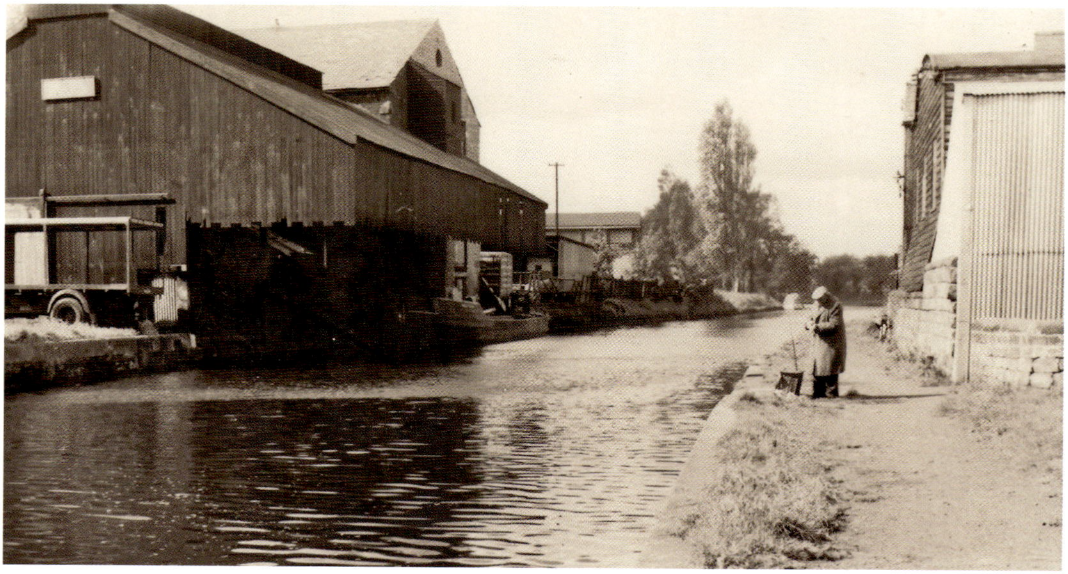

Stockton Heath Wharf, Bridgewater Canal

The wharf at Stockton Heath served Warrington and district, and was an important carriers' destination. Craft started to serve this place by September 1775, when a packet boat service also commenced. (*Railway & Canal Historical Society, Hugh Compton Collection, P. Norton 64274*)

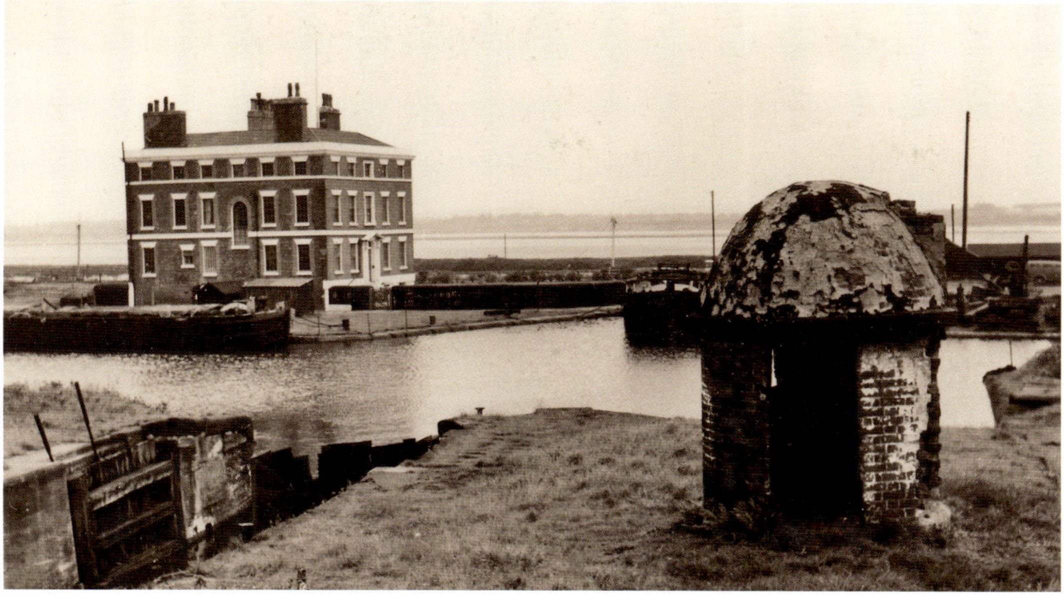

Bridgewater House, Basin and Locks, Runcorn, Bridgewater Canal, 1957
While the original flight became disused, the basin that linked the old line with the new and Runcorn Docks remained until the new lock route was also closed.

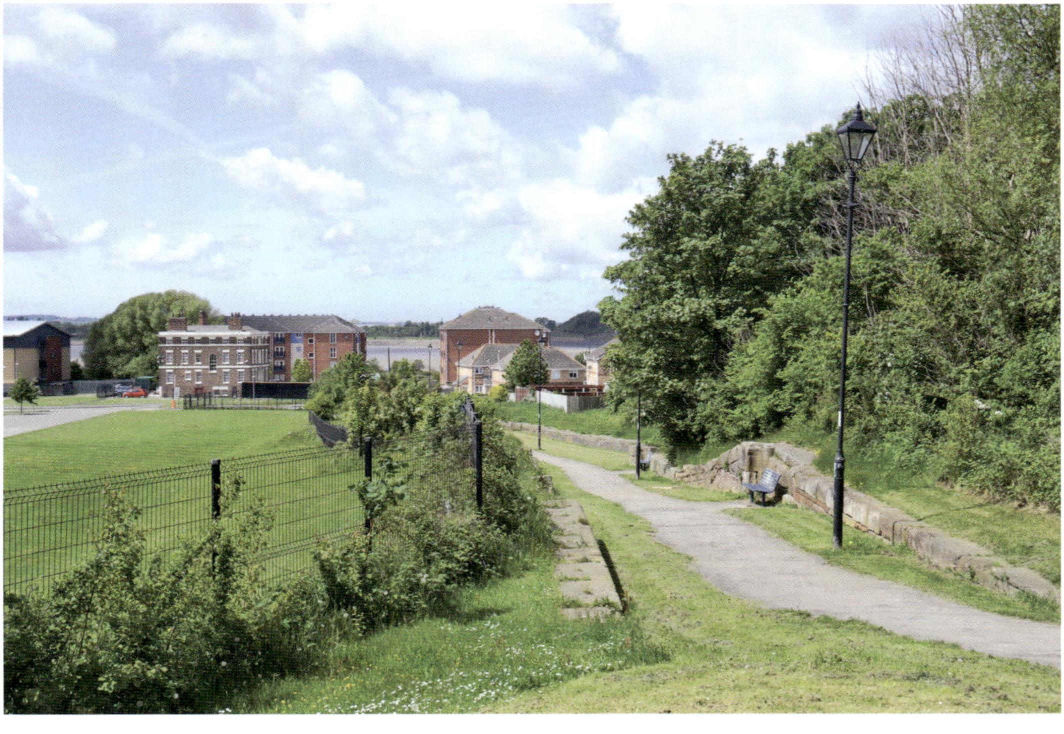

Bridgewater House, Basin and Locks, Runcorn, Bridgewater Canal, 2013
Part of the old staircase pairs of locks now form part of a walkway to new apartments, while the college now occupies the infilled basin in front of Bridgewater House.

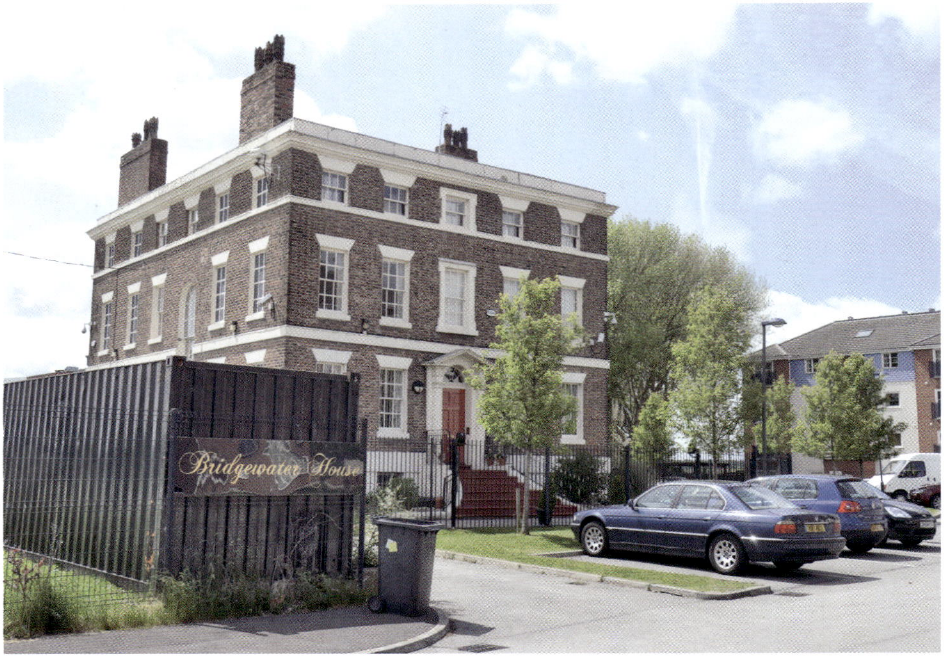

Bridgewater House, 2013
The old house as it is today. (*Ray Shill*)

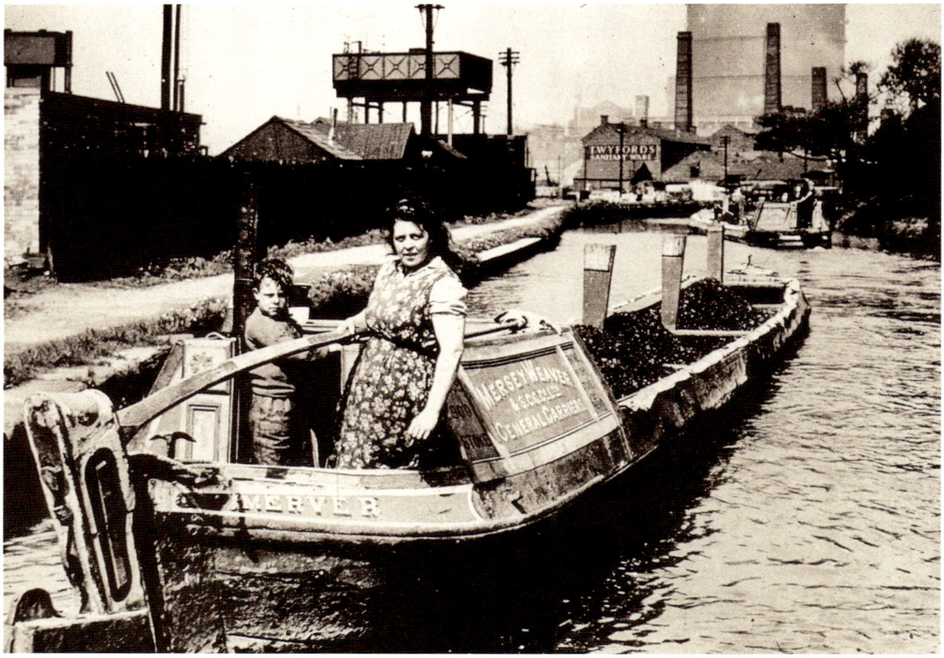

Runcorn Branch, Bridgewater Canal
The canal to Runcorn was the Mersey end of the main line from Manchester, which meant flats could lock down to the Mersey and cross to the Port of Liverpool where the Duke of Bridgewater had a dock. (*Railway & Canal Historical Society, Slide Collection 54503*)

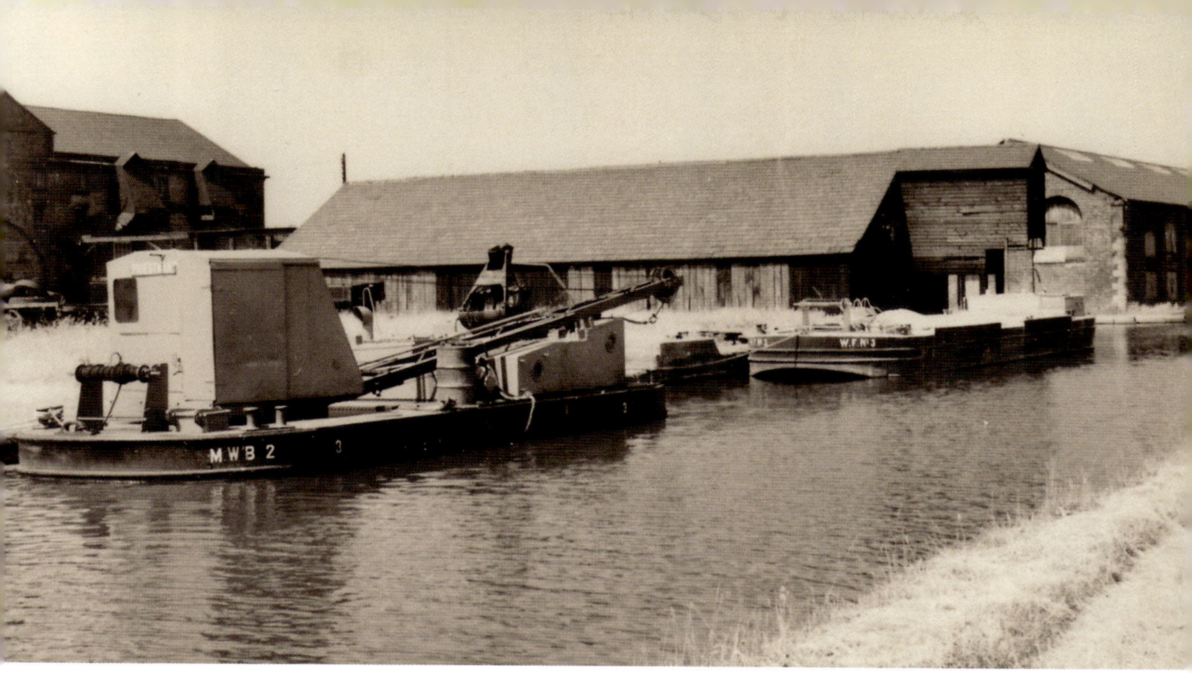

Preston Brook Warehouses, Bridgewater Canal, 1959 (*Above*) & 1961 (*Below*)
Preston Brook was a branch from the Bridgewater Canal, which joined the Trent & Mersey Canal. The warehouses established there became a major trans-shipment point for merchandise carried there by narrowboat. (*Railway & Canal Historical Society, Hugh Compton Collection 64264 & 64284*)

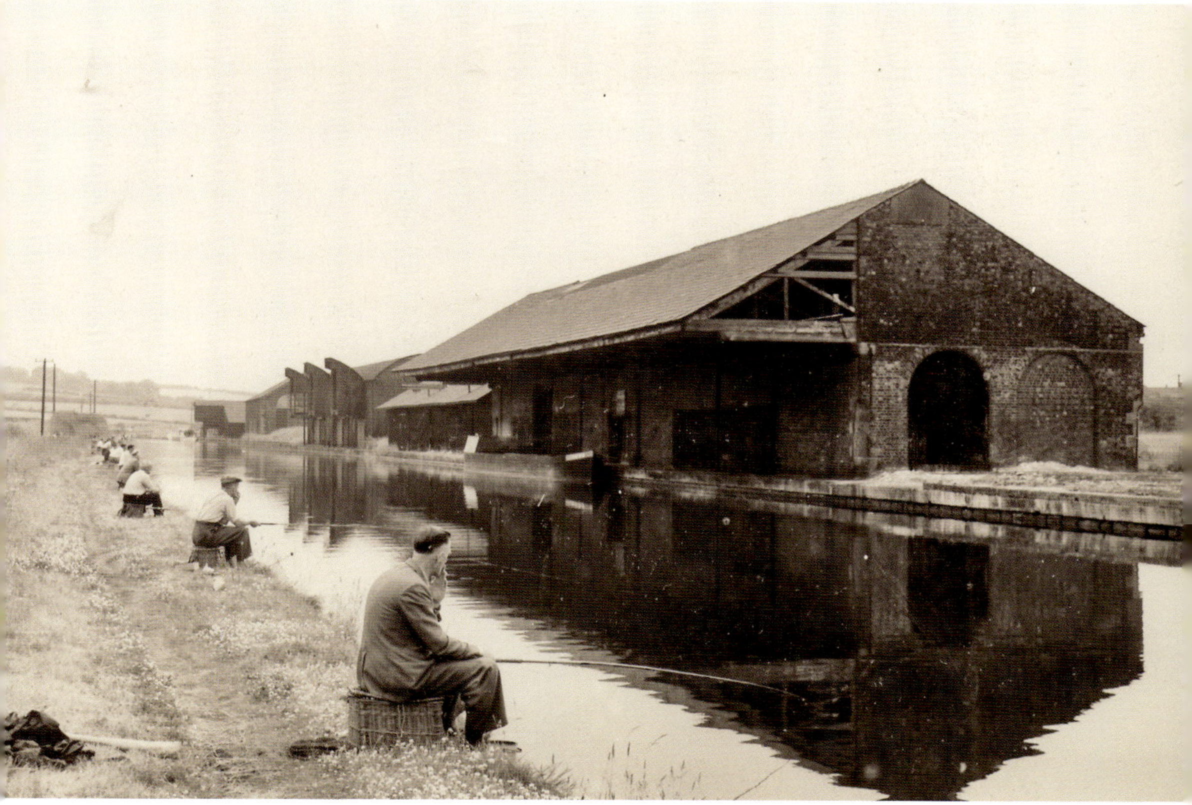

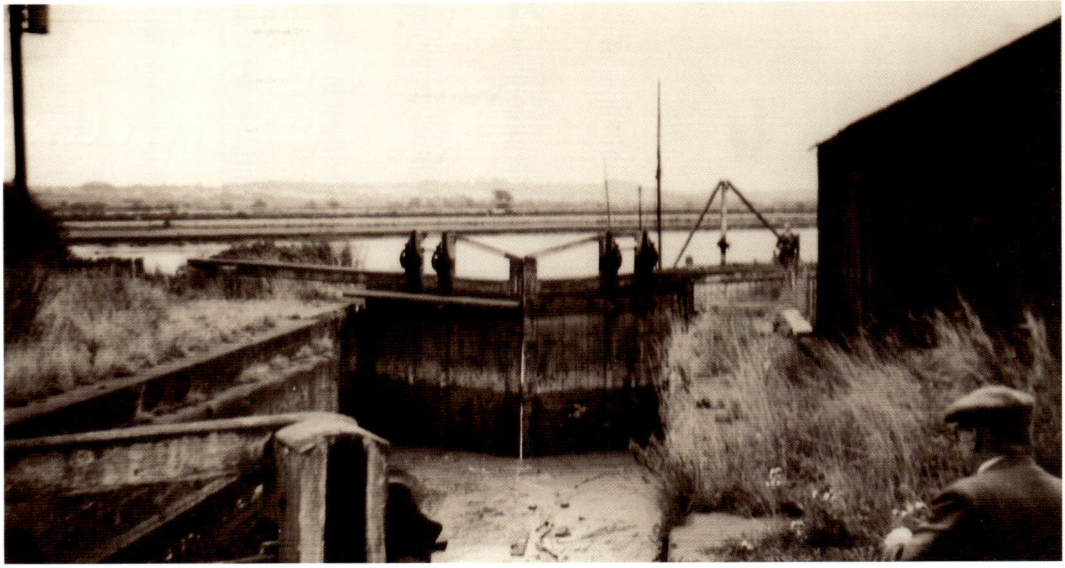

Fiddler's Ferry Lock, Sankey Brook Navigation

Initially, the Sankey Canal was linked to the navigable part of Sankey Brook which was tidal. The lock at Fiddler's Ferry was a downstream entrance to the Mersey, which was made to improve links with the river where boats could pass with reduced tidal restrictions. The extension from Sankey to Fiddler's Ferry was made under the extension act of 1762. During the 1980s, Warrington Council restored this lock for access to a new marina which had been formed out of a length of the canal. (*RCHS Hugh Compton Collection 65421*)

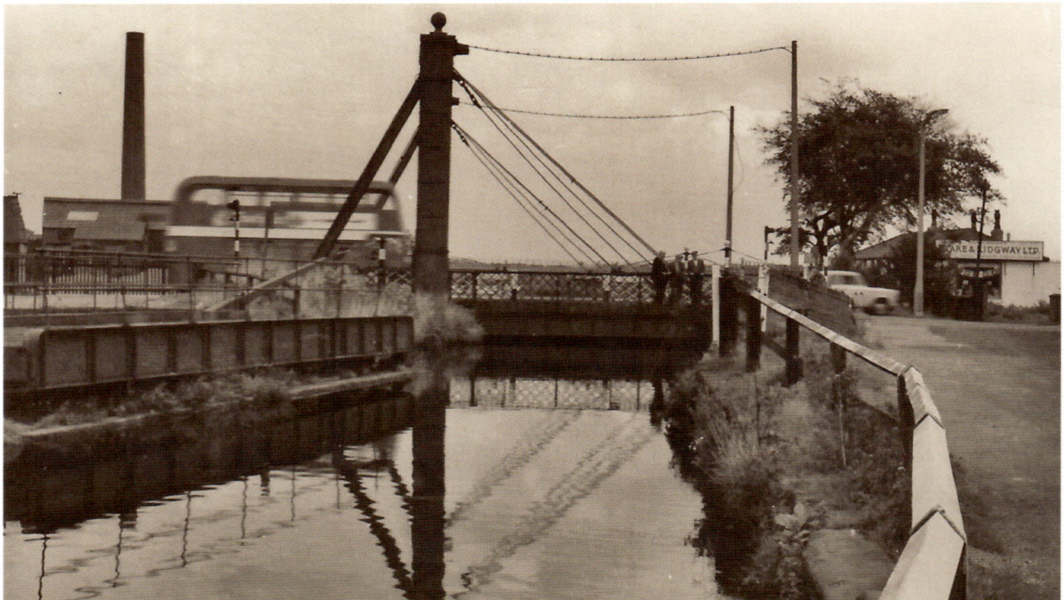

Road Lift Bridge, Warrington, Sankey Brook Navigation

The road from Warrington through to Widnes, originally a turnpike, crossed the Sankey Canal by a lift bridge. (*RCHS Hugh Compton Collection 65422*)

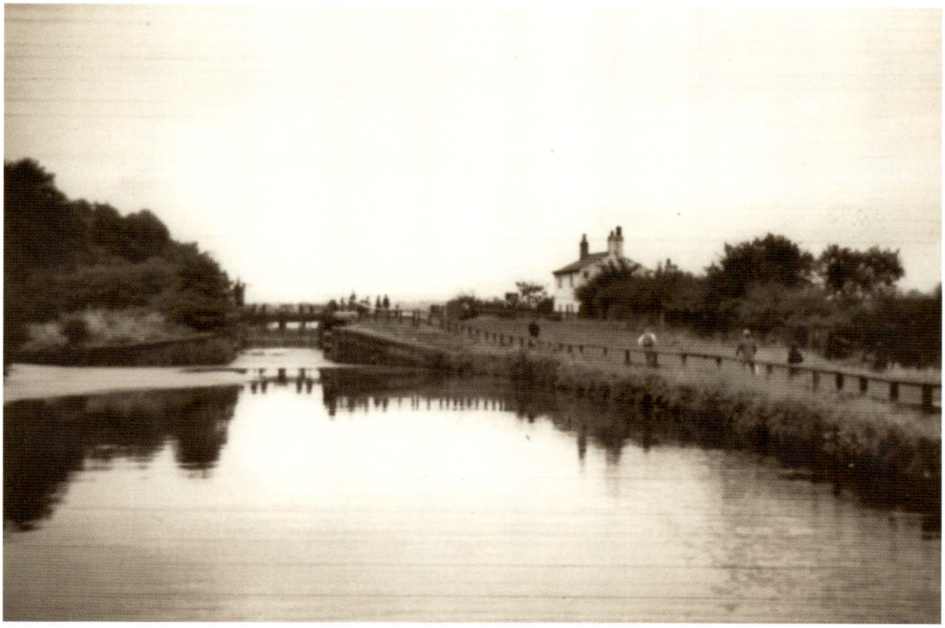

Bewsey Lock & Bewsey Lock House, Sankey Brook Navigation
Bewsey Lock was the first lock on the navigation after the river lock at Sankey Brook. Beyond are a further nine, which brought the waterway up to the summit level at St Helens. The last two locks were combined as a staircase pair, known as Double Lock. (*RCHS Hugh Compton Collection 65433 & 65436*)

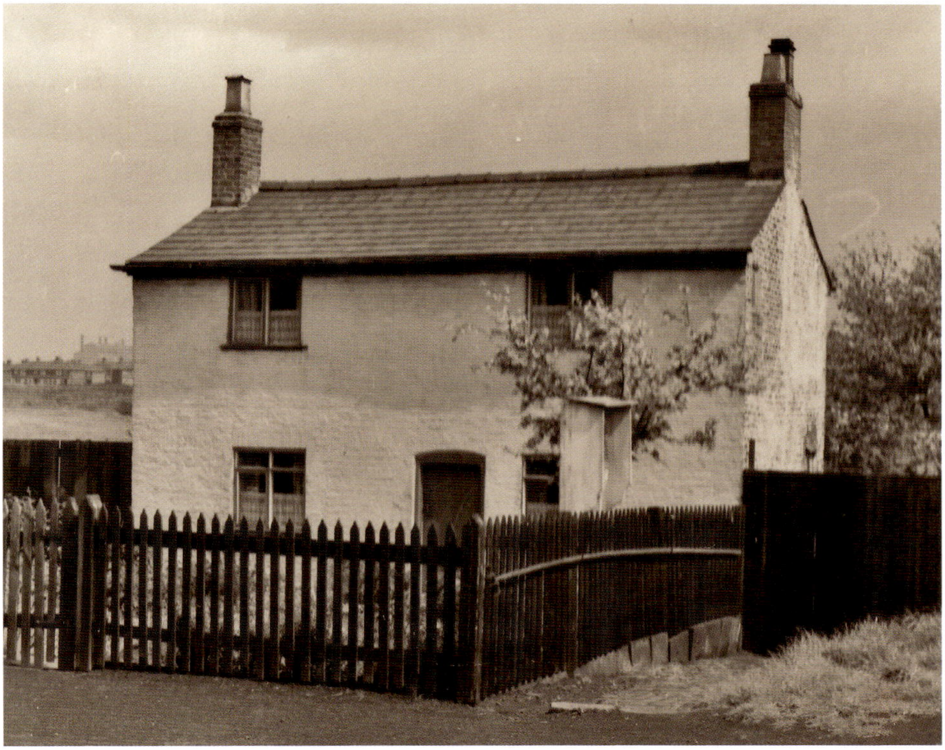

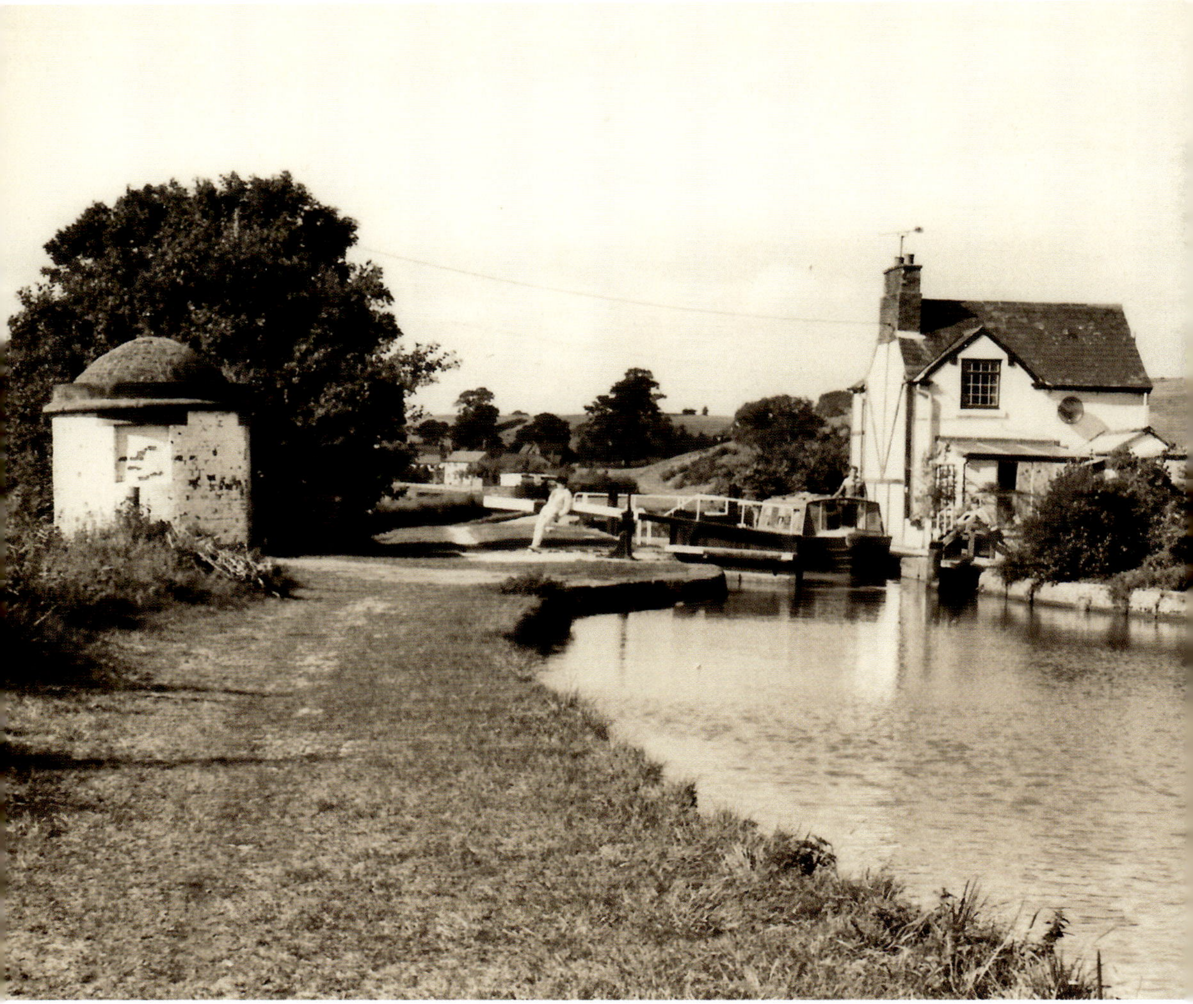

Beeston Stone Lock, Chester Canal

The Chester Canal was opened in 1779. Several features survive from the early times of the navigation to Nantwich, which include lock cottages and lengthmen's huts. (*Ray Shill*)

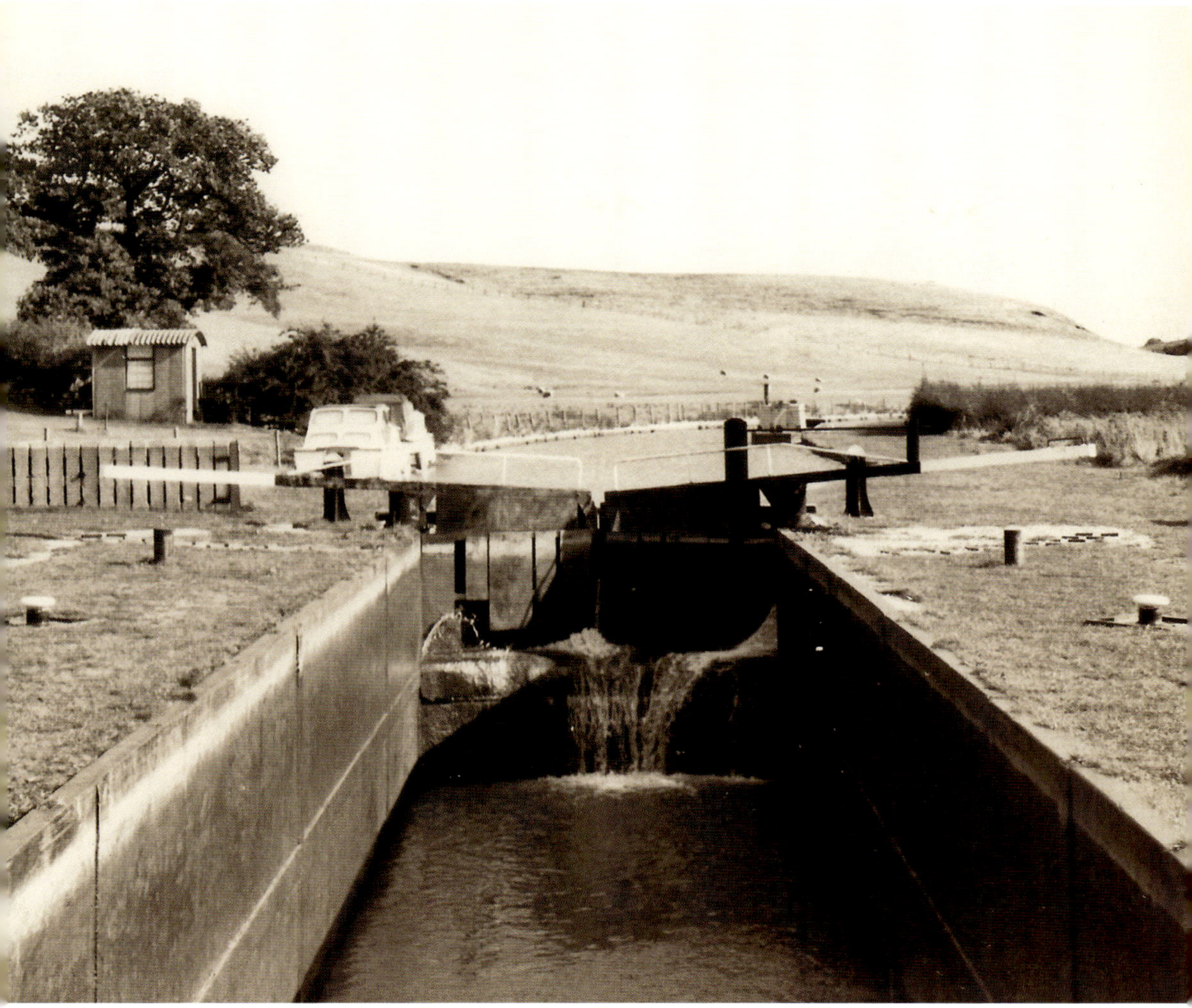

Beeston Iron Lock, Ellesmere & Chester Canal

The Ellesmere Canal and Chester Canal merged together in 1813 to provide a single navigation company from Chester into Wales. One of the locks at Beeston required rebuilding from time to time. The Ellesmere & Chester eventually decided on a novel solution and, guided by Thomas Telford's instructions, arranged to alter the lock during 1828. The cast-iron sides of this lock were provided to stabilise the ground around the lock, which was composed in part of running sand. The lock is Grade II listed by English Heritage. (*Ray Shill*)

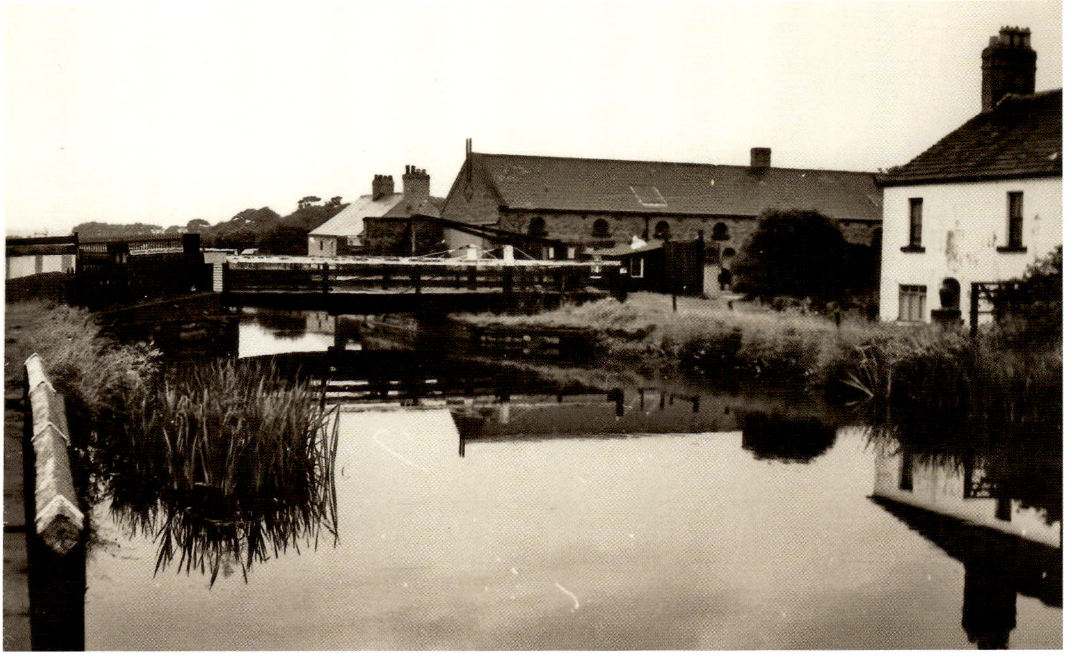

Winwick Quay, Warrington, Sankey Brook Navigation

The workshops and boat dock for the Sankey Canal were located at Winwick Quay, which served Warrington. (*RCHS Hugh Compton Collection 65439*)

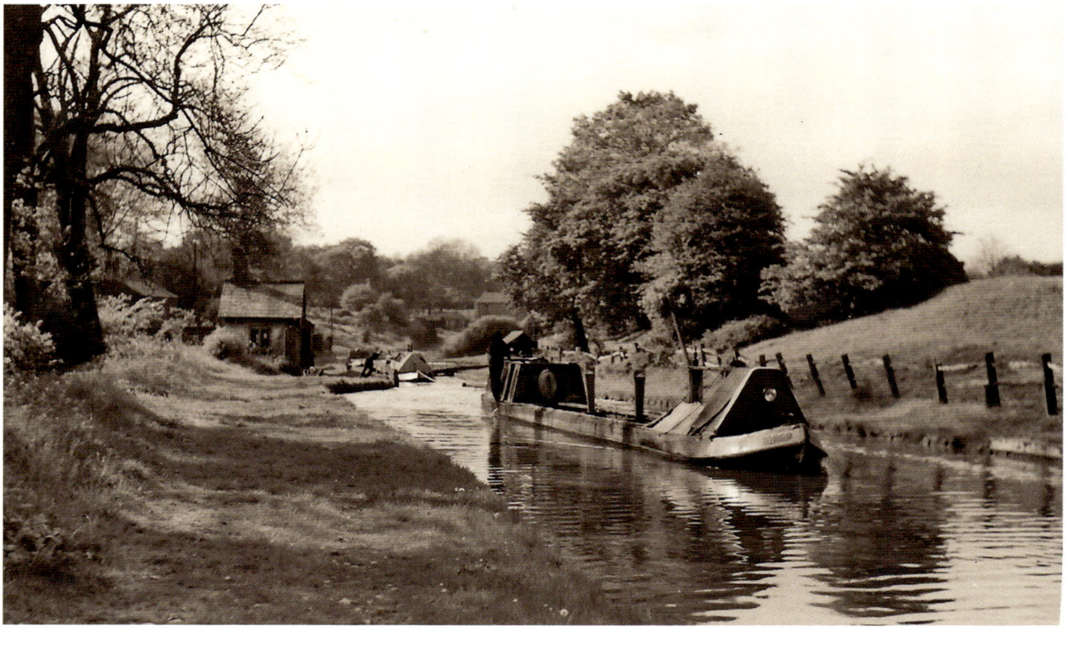

Dutton Stop, Trent & Mersey Canal

The line of the canal east of Preston Brook Tunnel came first to Dutton Stop, where there was a slight rise, before craft could pass onto the Trent & Mersey Canal. (*RCHS Hugh Compton Collection 65774*)

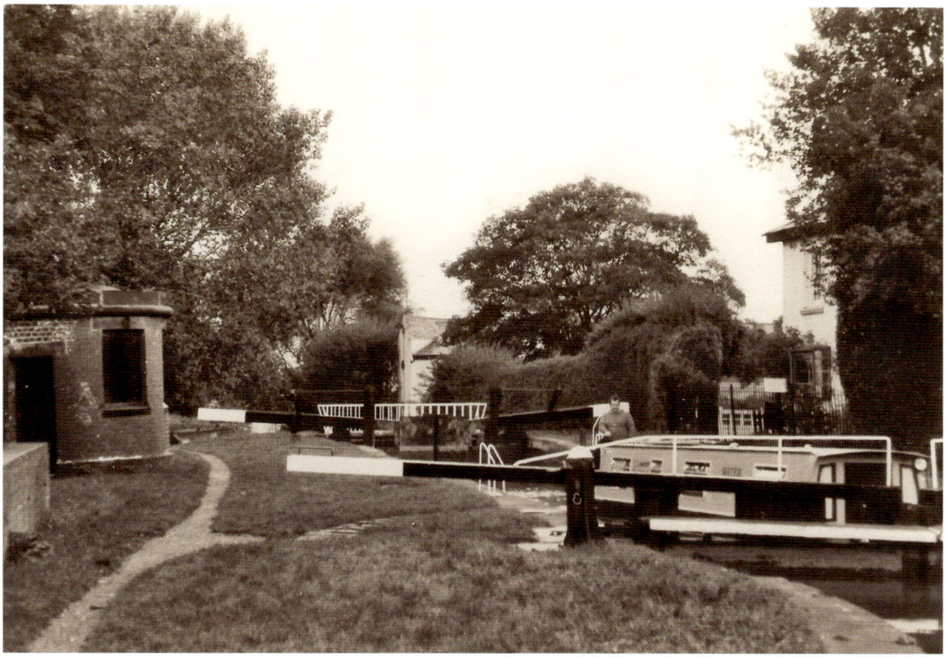

Tarvin Lock, Chester Canal

Tarvin Lock (now lock thirty-eight) has various features, including a lengthsman's hut and lock cottage. (*Ray Shill*)

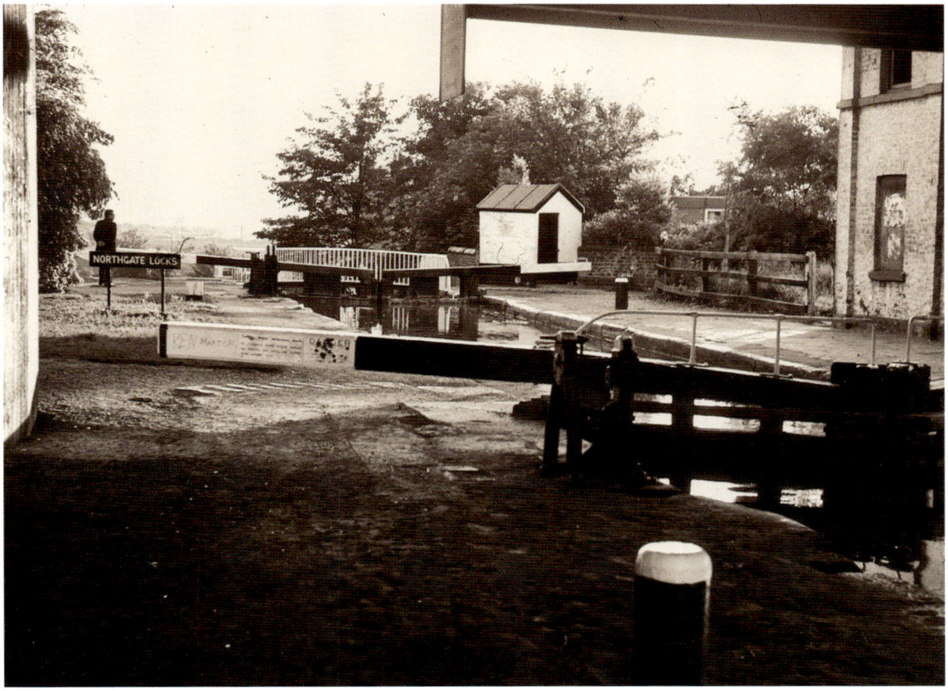

Northgate Locks, Chester Canal

(*Railway & Canal Historical Society 75171*)

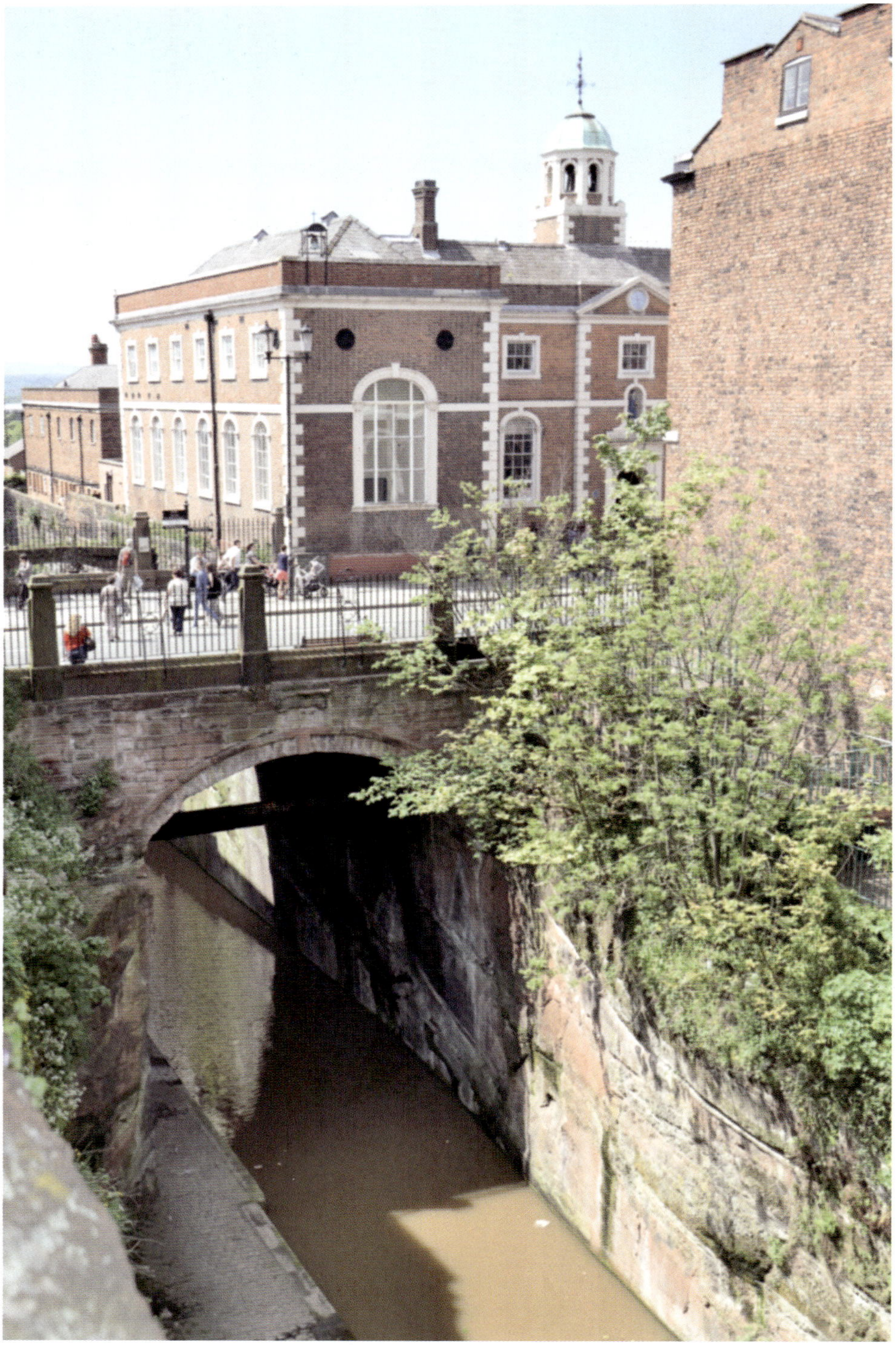

Northgate Bridge, Chester Canal

The Chester Canal was cut through rock close to the city walls and passed under Northgate in a deep cutting. In making this cut, the expense sustained was clearly detrimental to the company finances. (*Ray Shill*)

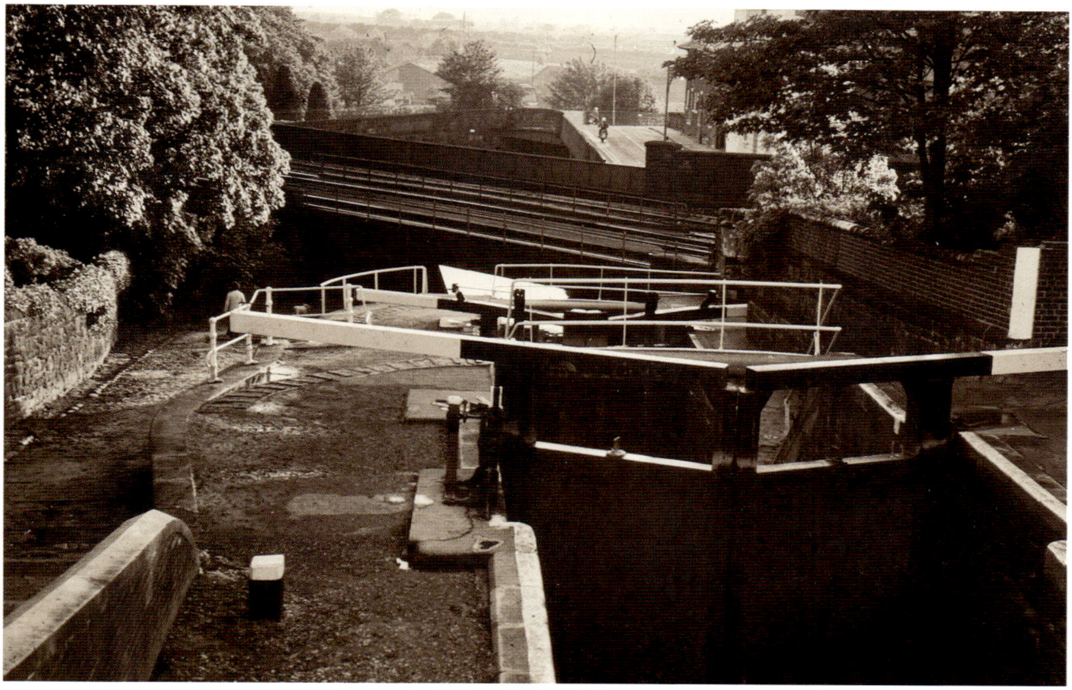

Northgate Locks, Chester Canal

Northgate locks originally comprised five locks, which it is thought were a staircase set of five barge locks; the lowest lock joined with the River Dee Tidal Basin. From 1795, this group of locks became a triple staircase and enabled boats to rise up from the junction with the Ellesmere Canal, within the confines of the steep rise required there, and then follow the city walls through the town at times in a deep cutting. (*Above: Railway & Canal Historical Society 75170. Below: Ray Shill*)

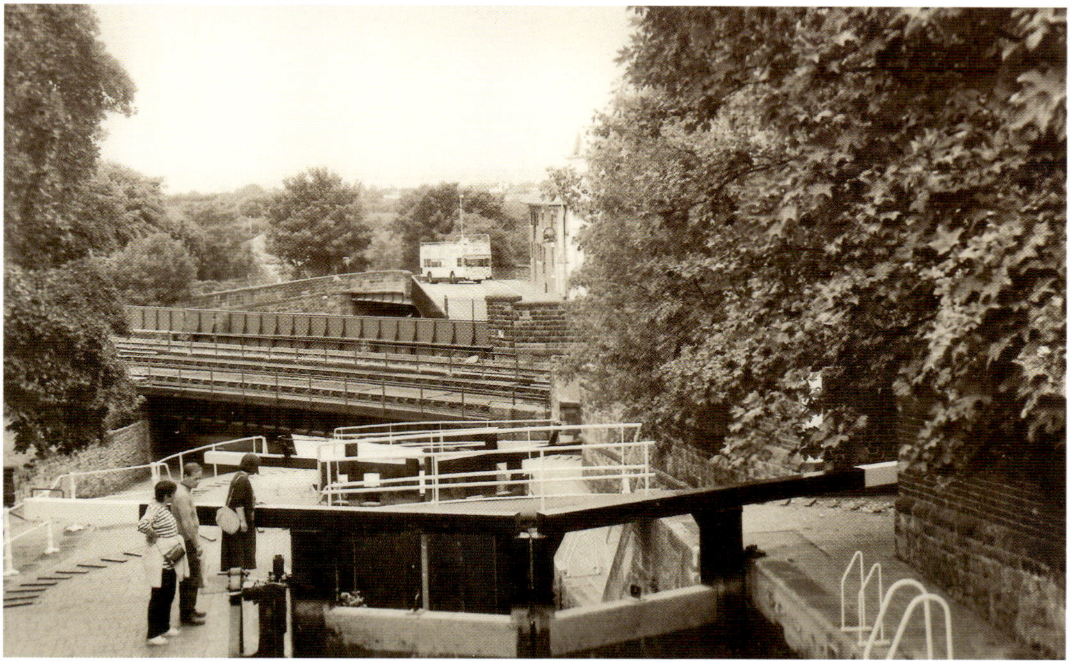

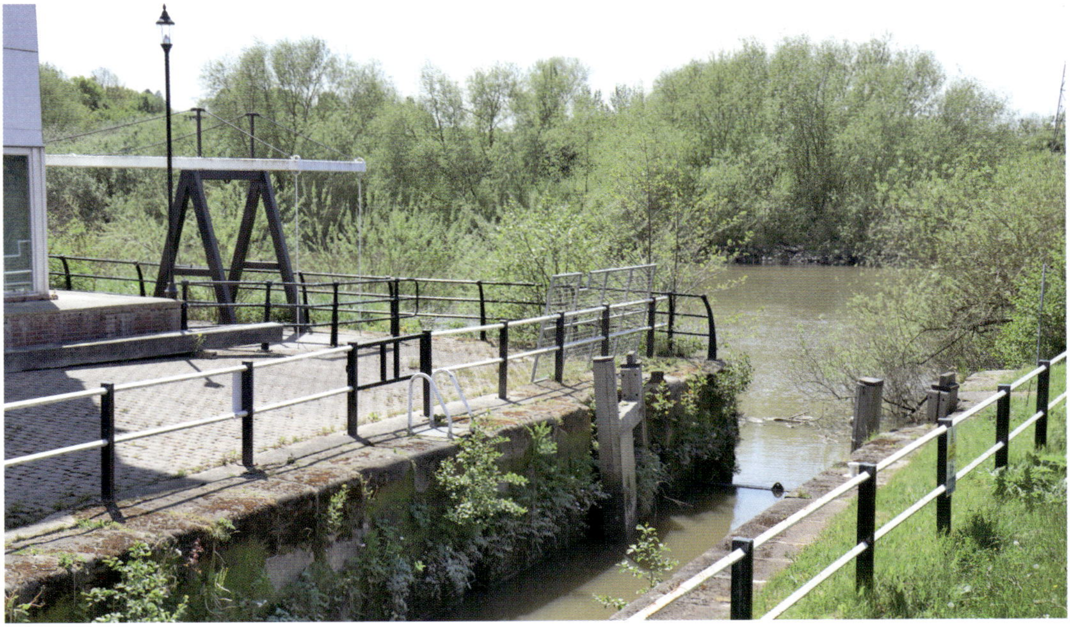

River Lock & Top Lock, Dee Branch, Chester Canal
There was at first a single set of gates that admitted vessels into the wide tidal basin beyond. With the completion of the Ellesmere Canal (1795), a lock was made for vessels, such as flats, into the basin that linked through another two locks to the Ellesmere Canal and Tower Wharf. The uppermost lock brought craft up to the main canal. From 1805, placed in between the two was a boat and graving dock. (*Above: Ray Shill*)

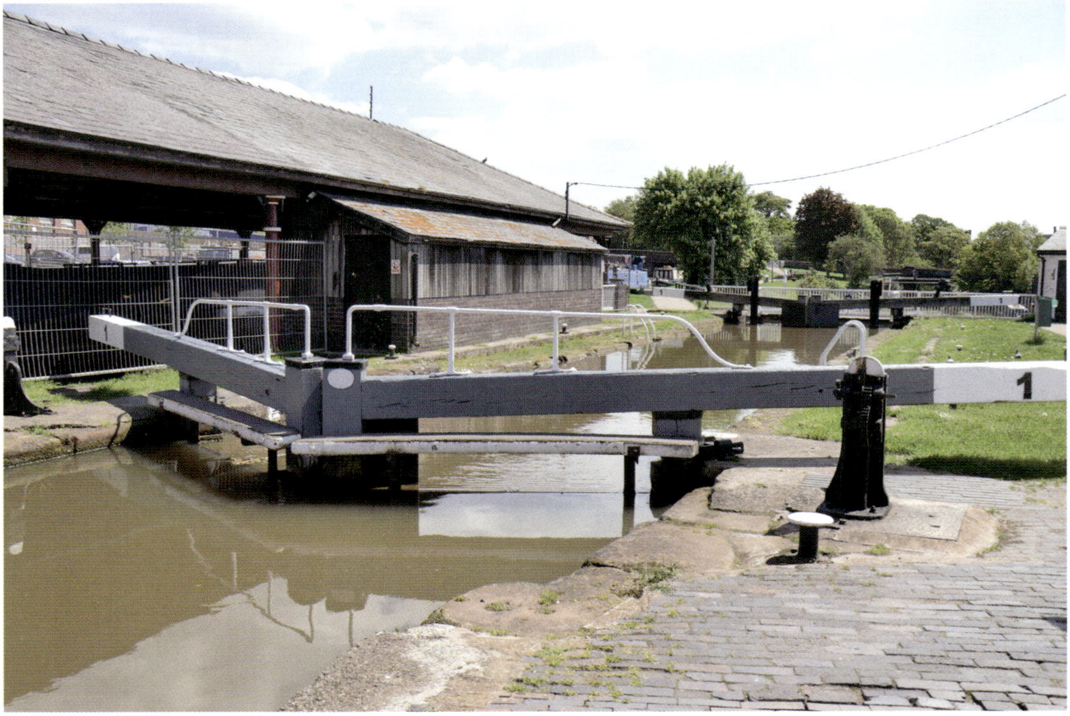

Chapter Three

Linking Canals & Rivers

Ellesmere & Chester (Chester–Ellesmere), Ellesmere & Chester (Middlewich Branch), Mersey & Irwell Navigation New Cuts (Runcorn–Latchford & Woolston Cut), New lock flight Runcorn (Bridgewater Canal), Runcorn and Weston canal and Sankey (St Helens) Canal extension to Widnes

Once the initial group of canals had been made, fresh schemes were put forward for other waterways. The Ellesmere Canal was a waterway with impressive engineering features; the line extended into Wales from the Chester Canal at Hurleston Junction as well as forming a new link from Chester to the River Mersey at Whitby.

With the opening of the section from Chester to Whitby (1795), a navigation fit for flats up to 14 feet wide could pass from the Dee through to the Mersey (and the navigation to Manchester), thus avoiding the coastal route and the danger inherent in navigating coastal waters as well as the sometimes turbulent waters of the 'Mersey Bar'.

In making this navigation, the junction of the River Dee and Chester Canal was altered, with two new locks ascending to the level of the Ellesmere Canal opposite their main Chester Wharf Tower. The Ellesmere Canal was also extended to join up woth the Chester Canal between the second and third lock. The lowest two barge locks of the Chester Canal were then abandoned.

The locks at Whitby numbered three barge locks. As a settlement developed, this place became known as Ellesmere Port. Much of the Ellesmere Canal was completed by 1805, although a section of the intended waterway near Wrexham was never built. In 1813, the Ellesmere Canal Company merged with the Chester Canal Company to form the Ellesmere & Chester Canal Company.

Ongoing improvements to the Mersey & Irwell Navigation included the making of longer lock cuts. In 1804, the Runcorn & Latchford Canal was opened from the Mersey at Manor (Latchford) Lock, through to Runcorn, creating a new port and dock on the east side of the town. Access to the Mersey was achieved through a basin once suitable tide levels had been reached. A second adjacent, and larger, basin was later made here to hold more craft. In this way boats waiting for the tide might be accommodated for departure speedily, rather than waiting their turn in a queue to pass the bottom lock.

Water supply for the Runcorn & Latchford Canal was conserved by a feeder, which brought water from above the Old Woolston Cut river lock near Thelwall. During 1821, a second important lock cut was completed from Woolston to Paddington (the New Woolston Cut), and the feeder link was altered to leave the pound above Paddington lock and cross the Mersey by a wooden trough to Meadow Lock. Another scheme that included a ship canal, proposed in 1840, was not proceeded with.

Improvements to the rival Duke of Bridgewater's establishment at Runcorn was also made with the provision of new docks, while between 1825 and 1827 a second lock flight was added to enable a larger number of craft to pass to and from the river as trade increased. Later changes included an extension to the waterway at Runcorn, when opened in 1859. The link between the Runcorn Locks and the Weston Canal ensured they also had access to the port of Weston.

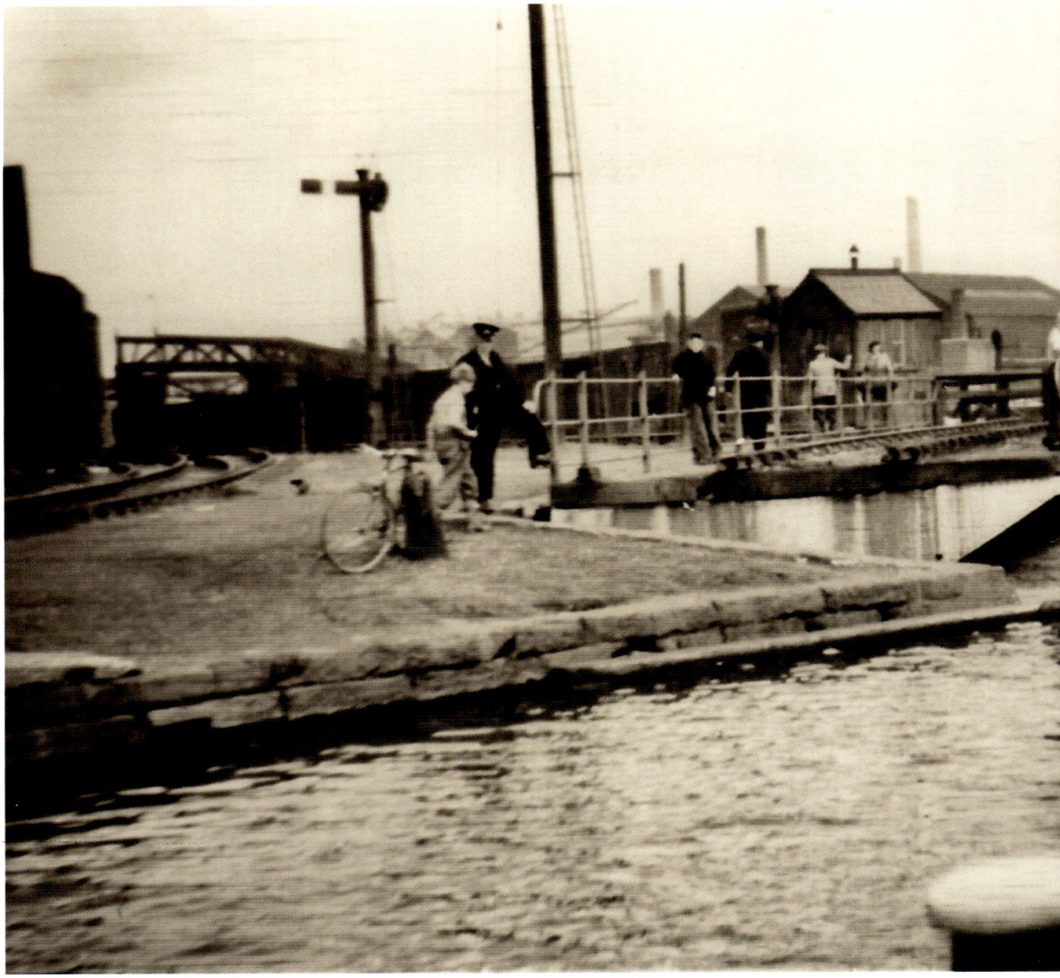

Railway Swing Bridge, Widnes, Sankey Brook Navigation
The Sankey Canal was extended to Widnes in 1833, further improving the communications to the Mersey, which hitherto had been Fiddler's Ferry. The St Helens Railway to Widnes Dock crossed the Sankey Canal by a swing bridge. (*RCHS Hugh Compton Collection 65422*)

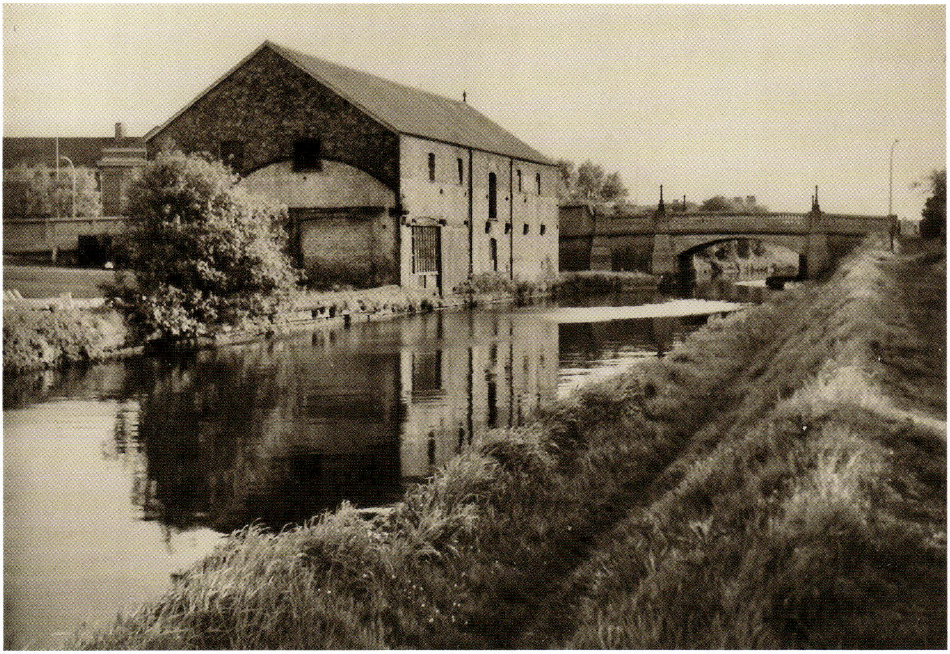

Black Bear Wharf and Knutsford Road Bridge, Runcorn & Latchford Canal, Mersey & Irwell Navigation

The section retained for navigation between the Manchester Ship Canal and the Mersey has been filled in and is now a walkway and cycle path. In the upper view, the canal when in use is seen, while the lower image is a modern view. (*Above: Railway & Canal Historical Society, Norton 65204. Below: Ray Shill*)

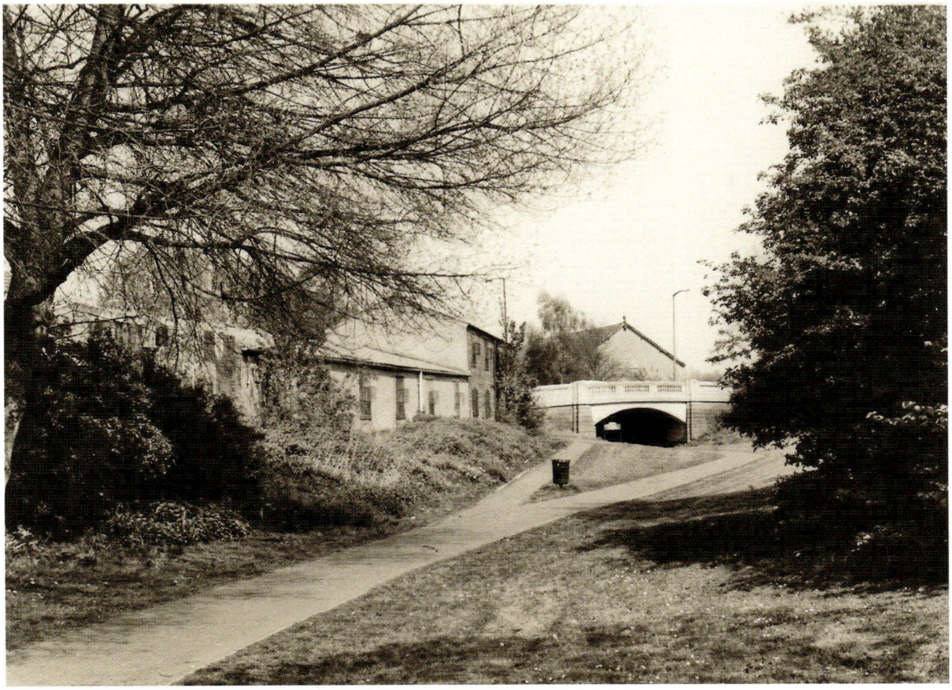

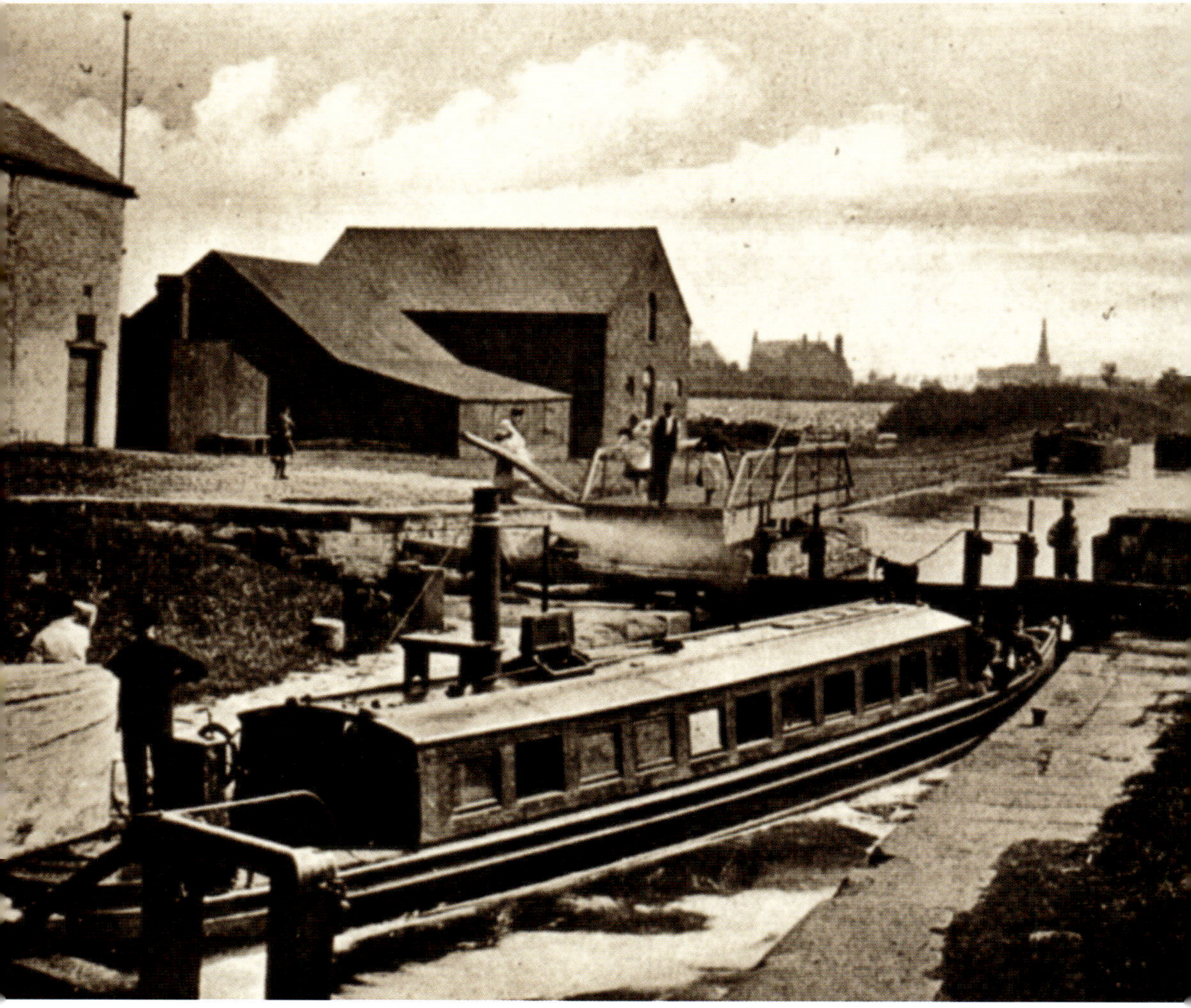

Manor (Latchford) Lock, Runcorn & Latchford Canal, Mersey & Irwell Navigation
The Runcorn & Latchford Canal was once a busy waterway that formed the link between Runcorn docks and the River Mersey at Latchford. The route was later bisected by the Manchester Ship Canal but the section from Twenty Steps Lock to Manor Lock was retained for craft passing onto the navigable part of the Mersey. (*Railway & Canal Historical Society, Slide Collection Ref 58818*)

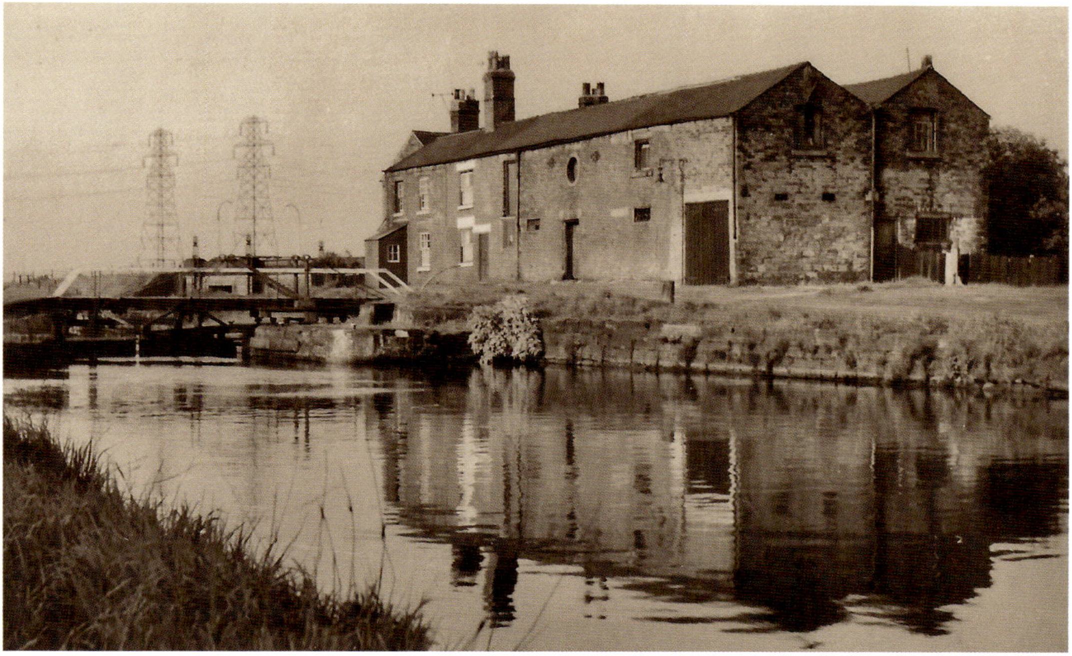

Manor Lock, Runcorn & Latchford Canal

The junction with the Mersey, now (*below*) and then, when the Runcorn & Latchford Canal was used for navigation (*above*). (*Above: Railway & Canal Historical Society, Norton 65206. Below: Ray Shill*)

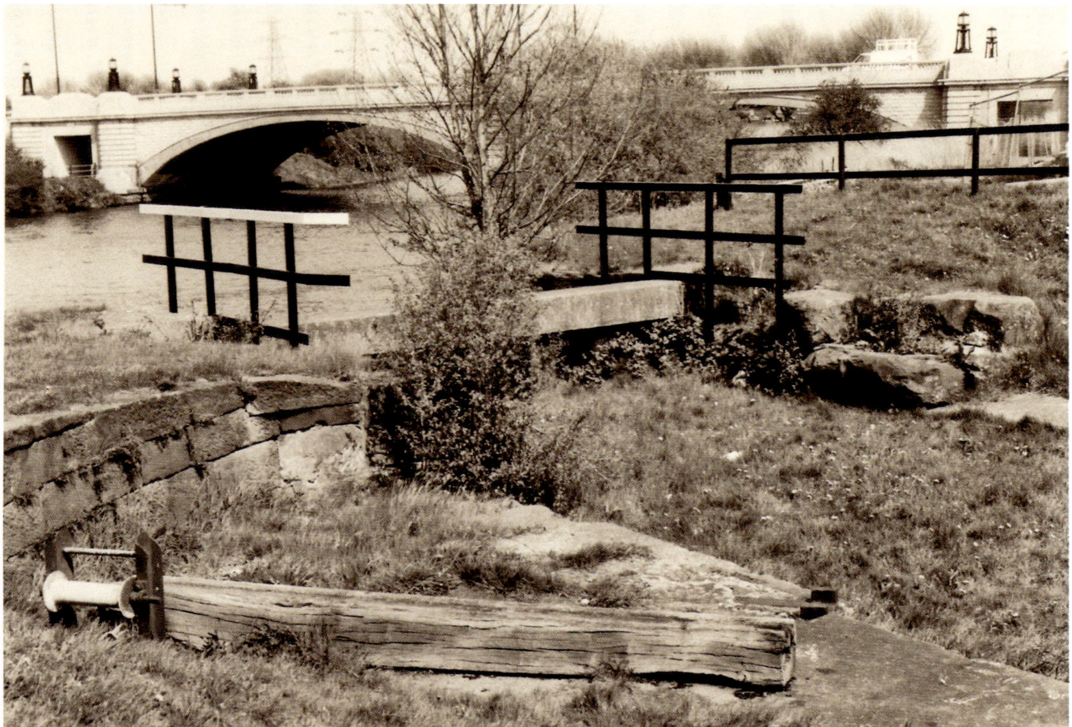

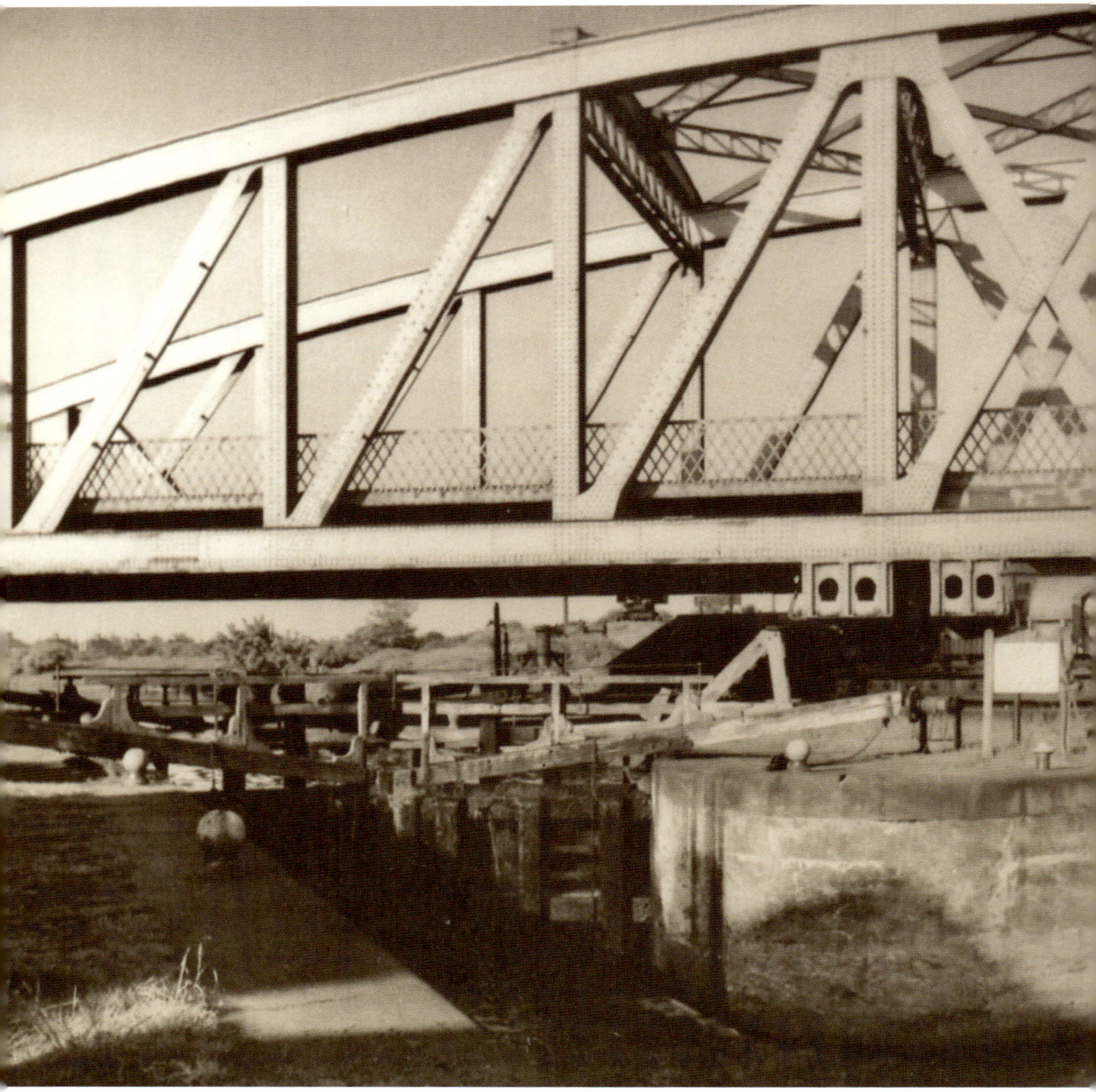

Twenty Steps Lock, Runcorn & Latchford Canal
This lock enabled craft to pass to and from the Manchester Ship Canal and was overshadowed by the massive swing bridge that carried the road over the Ship Canal to Sutton Heath. (*Railway & Canal Historical Society, Norton 65202*)

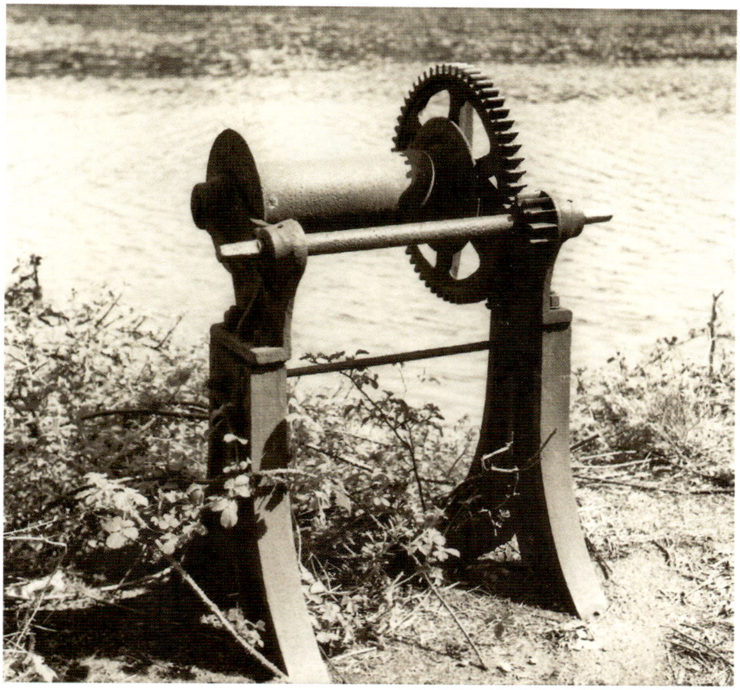

Winch, East End Woolston Cut, Mersey & Irwell Navigation
A survivor from the days of navigation is this winch. Such was the nature of river flow that accessing Woolston Lock, assistance might have been needed to pull craft into the lock chamber. (*Ray Shill*)

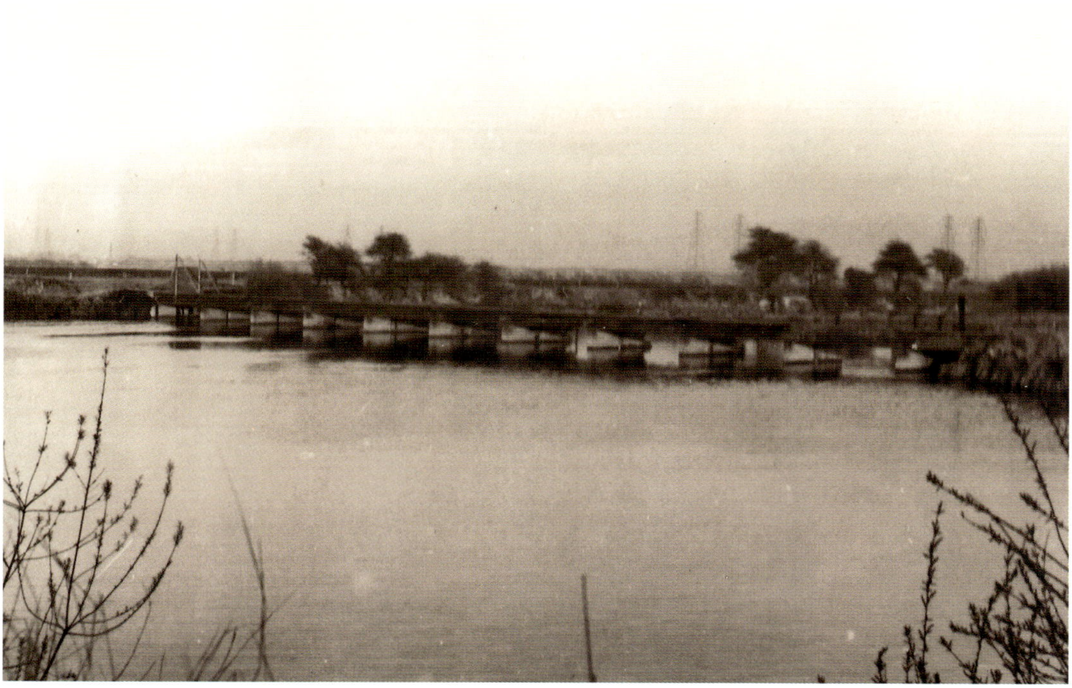

Downstream End, Woolston Cut, Site of Paddington Lock and Timber Trough, Mersey & Irwell Navigation
Woolston New Cut provided the water for the Latchford & Runcorn Canal using a timber trough that crossed the Mersey. The site of Paddington Lock was located on the far left. (*Railway & Canal Historical Society, Norton 65195*)

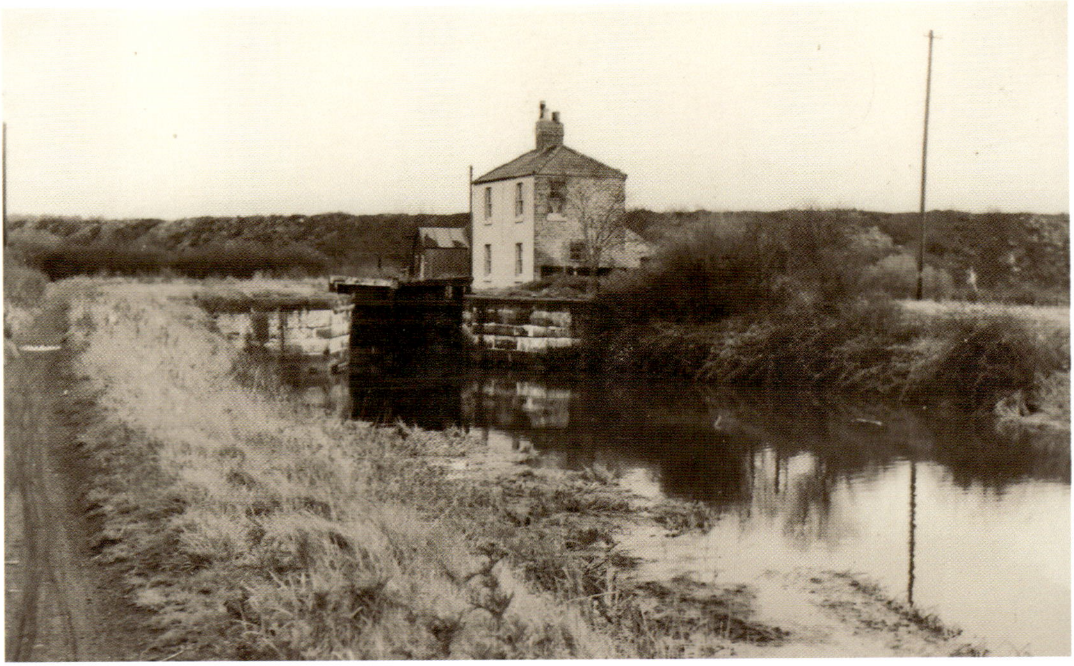

Upstream End, Woolston Cut & Lock House, Mersey & Irwell Navigation

The original lock house is seen when this canal had ceased to be used, but still supplied water to a Bone Works and the Runcorn and Latchford Cut. The lower image dates from 2006 after the building on the site of the lock house had been enlarged. (*Above: Railway & Canal Historical Society, Norton 65198. Below: Ray Shill*)

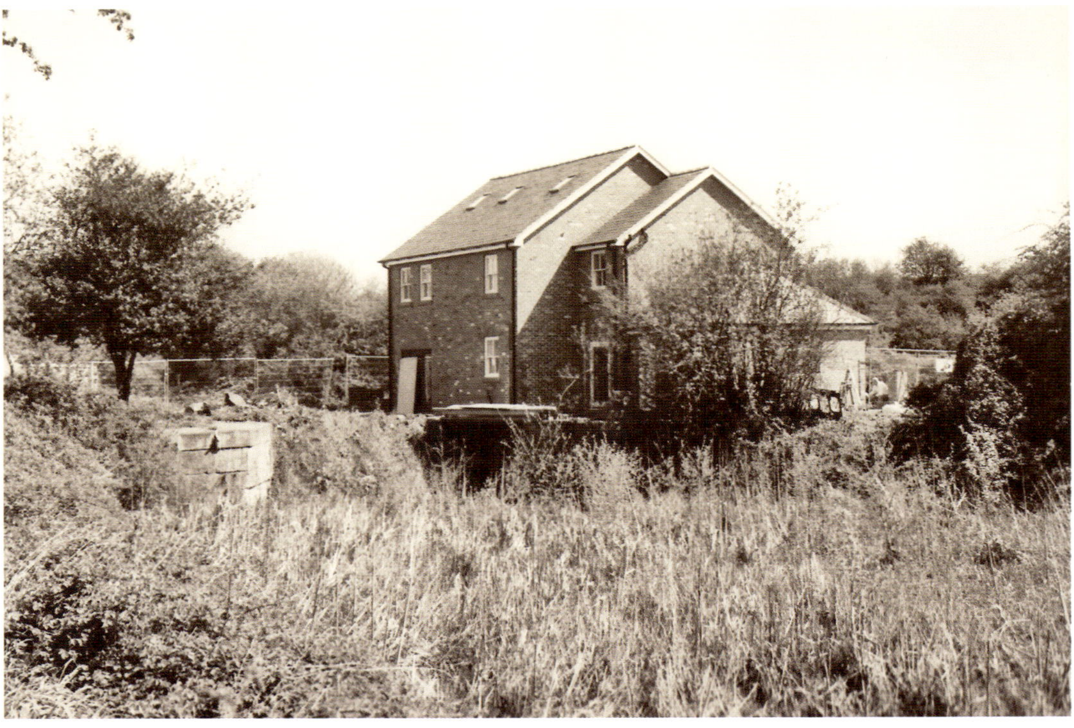

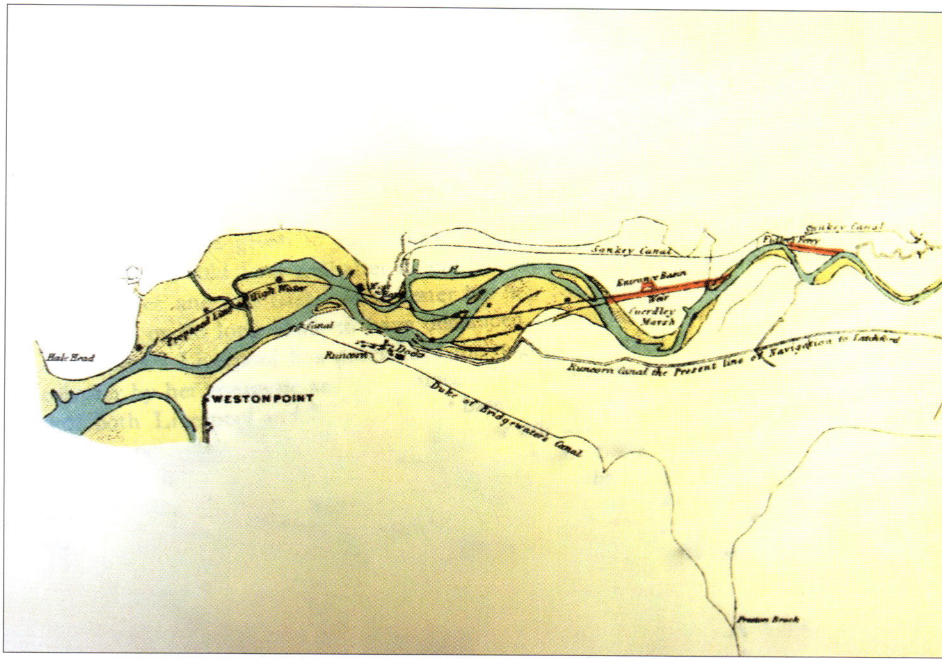

Henry Palmer's Plan for a Ship Canal from Weston Point to Manchester, 1840

Henry Palmer produced a survey for a group of proposed new cuts, which was not adopted. It is a useful plan in that it also shows the waterways that were made. (*Heartland Press Collection 786019 & 786020*)

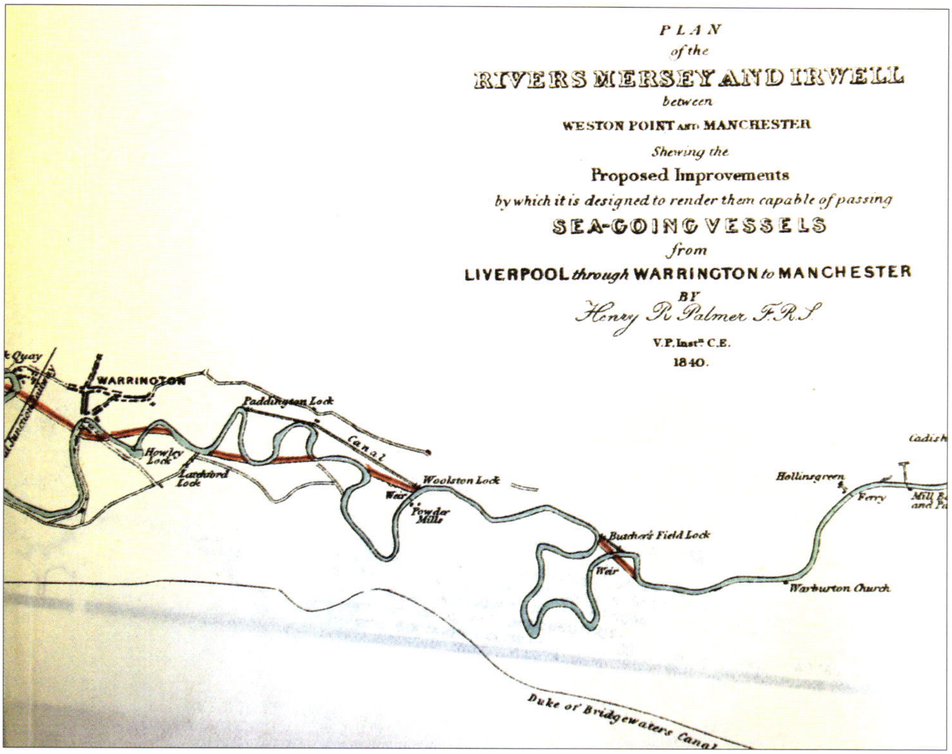

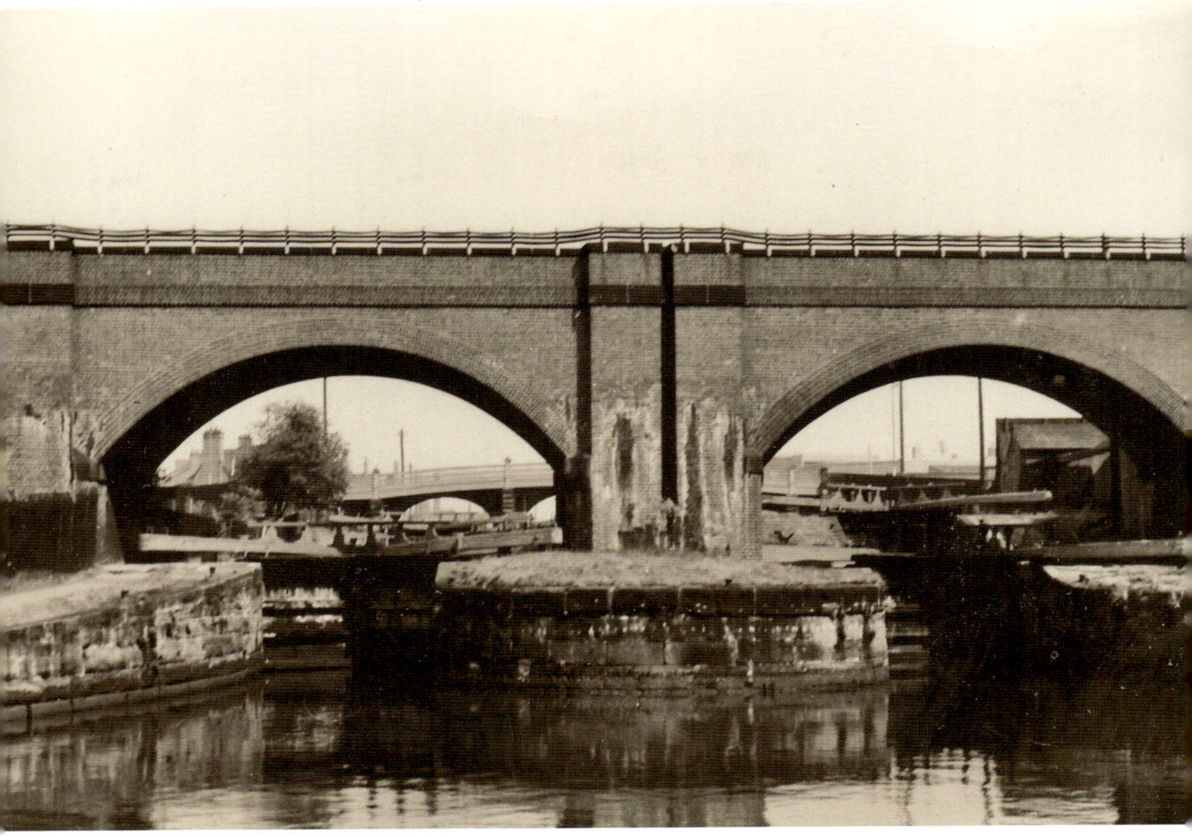

Runcorn Locks, Bridgewater Canal, 1957

In this image the angle of the view is through the railway arches to Waterloo Bridge. Originally Located here was a pair of staircase locks that descended to this point. These early locks were opened in January 1773 and formed part of five sets of paired locks that descended to the Mersey. During the 1820s, this arrangement was changed with the making of two sets of parallel single locks to this point. A new flight of single locks then diverged from the old route and continued on to a separate part of Runcorn Docks and the link to the Mersey. (*Railway & Canal Historical Society, Hugh Compton Collection, P. Norton 64291)*

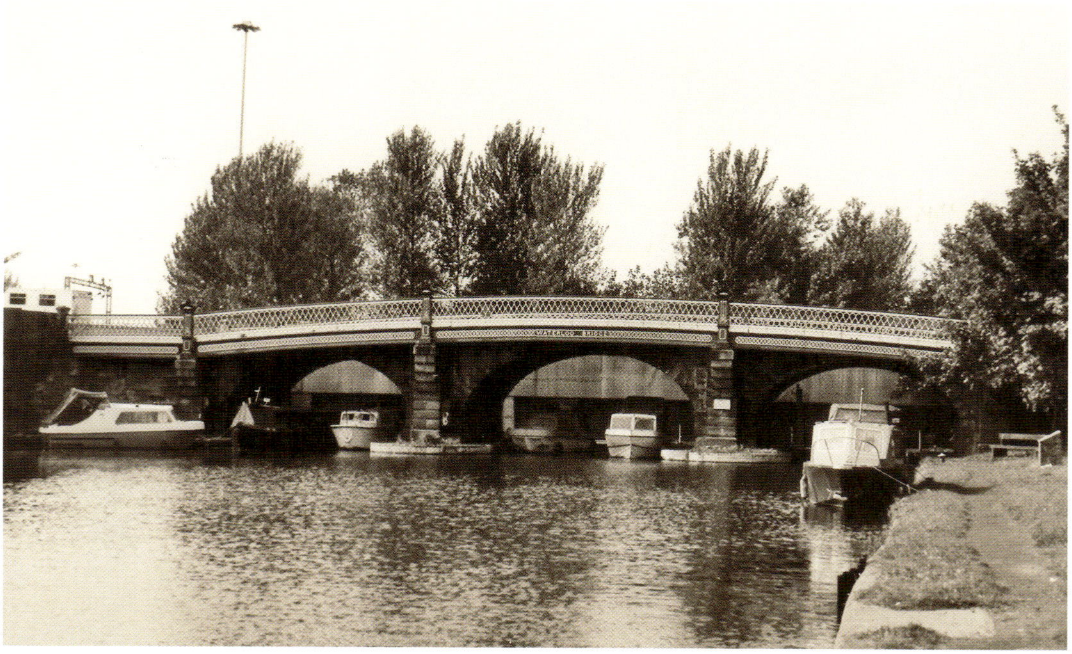

Waterloo Bridge, Runcorn Locks, Bridgewater Canal, 1998
The top locks of the ten flight set of locks to the Mersey were located on the west side of Waterloo Bridge. What is left of the lock entrances have been retained for moorings. (*Ray Shill*)

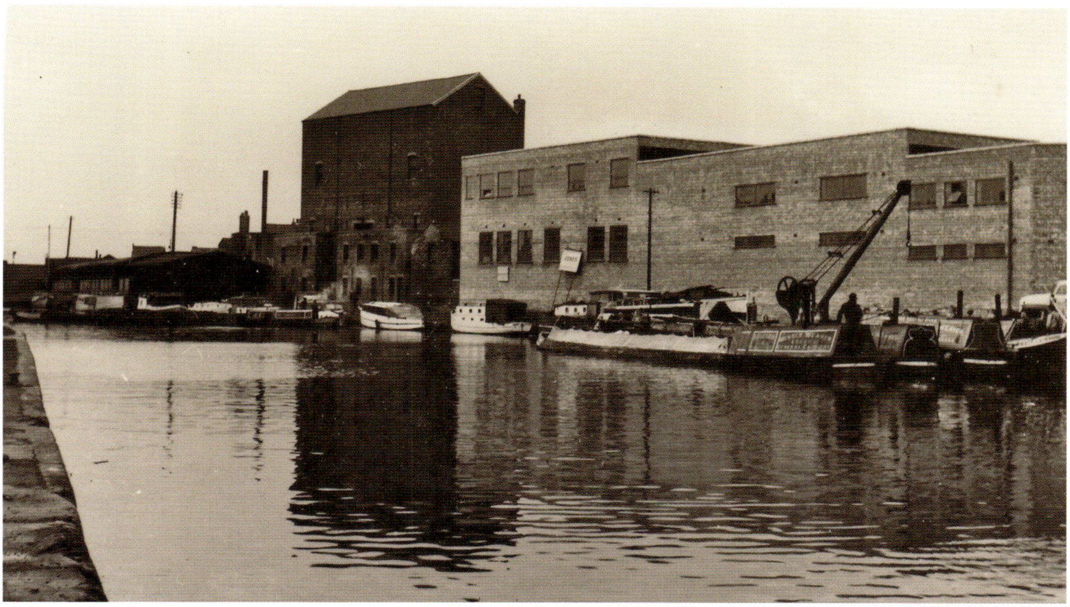

Runcorn Basin, Bridgewater Canal, 1957
The Bridgewater Trustees made various improvements to the docks at Runcorn, providing hydraulic power for cranes and developing docks such as Alfred, Fenton and Francis there. Francis Dock was a widened part of the Runcorn and Weston Canal. (*Railway & Canal Historical Society, Hugh Compton Collection, P. Norton 64289*)

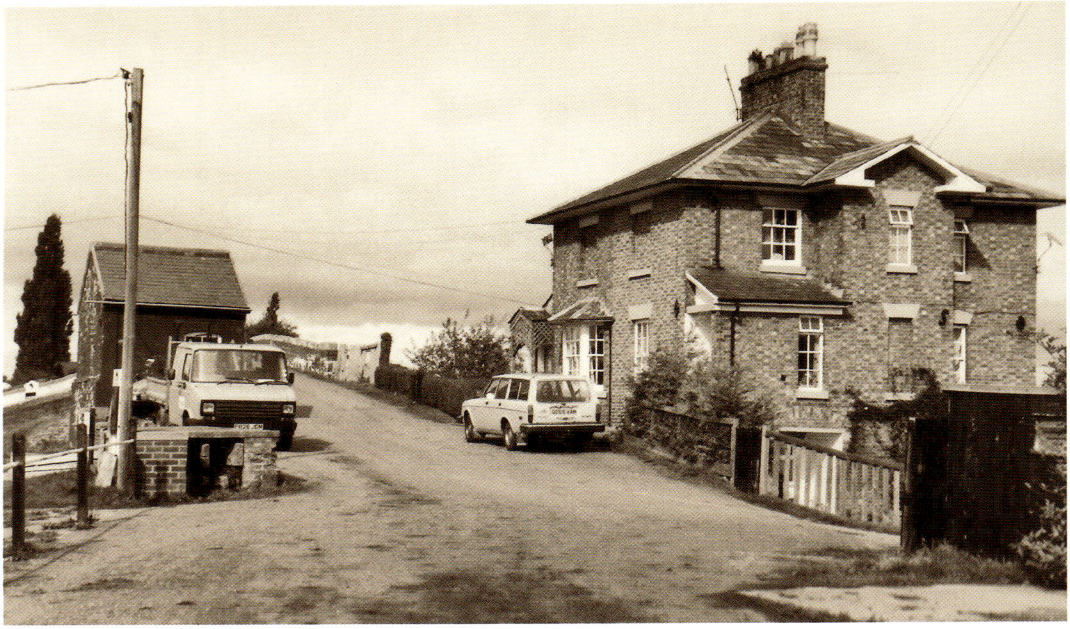

Canal Houses, Barbridge Junction, Shropshire Union Canal
Barbridge became the junction between the Middlewich Branch and the Chester Canal. The houses at Barbridge have features of being designed by Telford. There was an overall roof that spanned the main canal, but this was later removed. (*Ray Shill*)

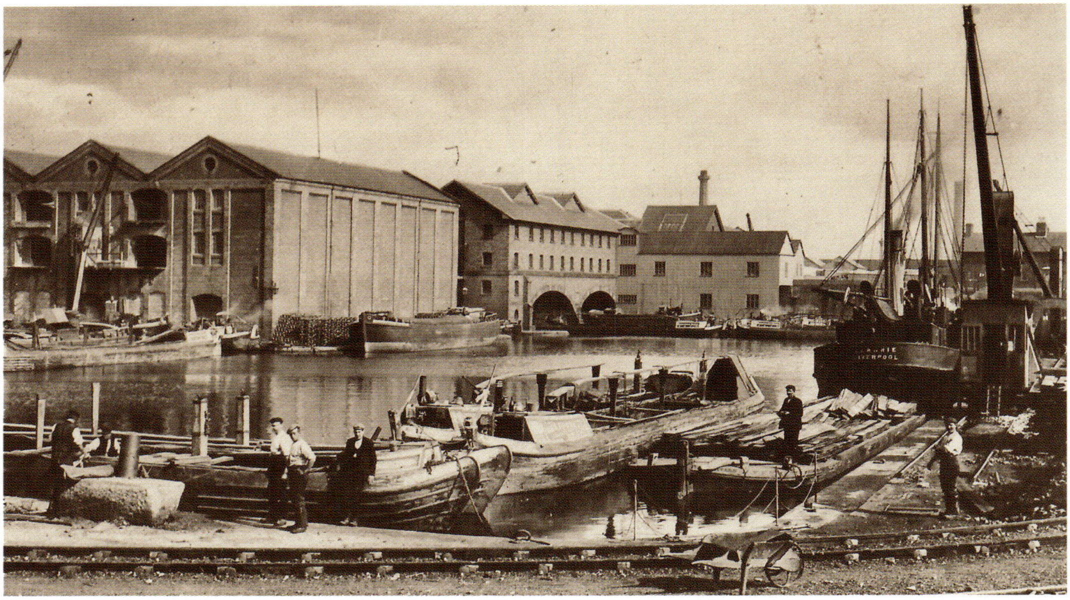

Ellesmere Port, Telford Warehouses, Lower Basin, Shropshire Union Canal
All these buildings were lost in a fire, but in their day were an important trans-shipment point for goods carried by narrowboat. They were designed by Thomas Telford and are a tribute to his engineering skill. While work on building these warehouses began when Telford was still alive the period each part was completed extended over ten years from 1833–43. (*John Ryan Collection*)

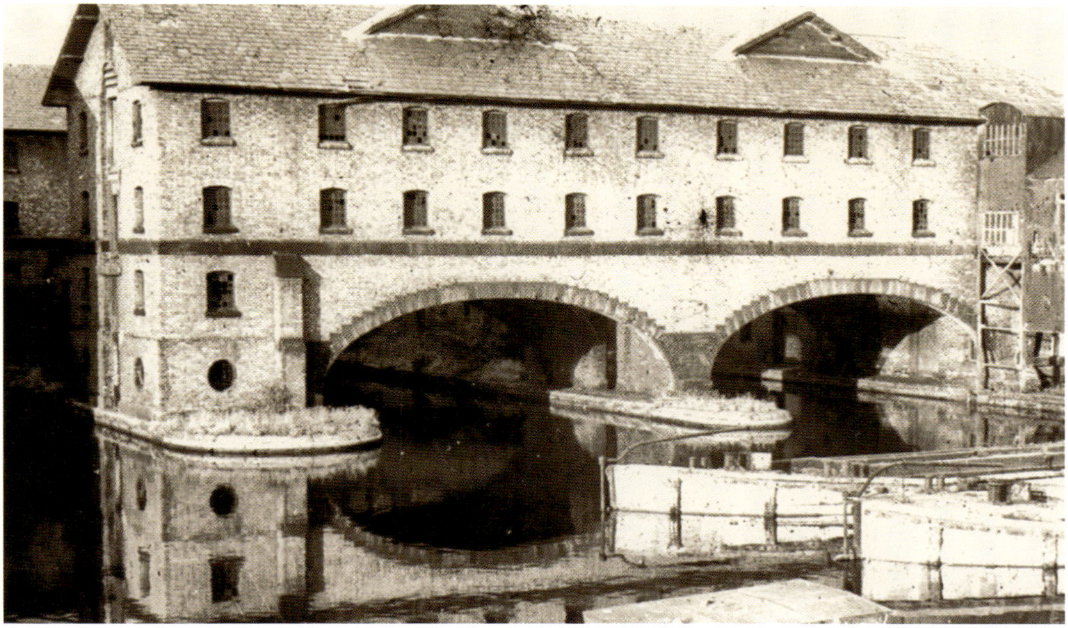

Ellesmere Port, Telford Warehouses, Lower Basin, Shropshire Union Canal
A feature of the Telford warehouses at Ellesmere Port included a part that spanned moorings in the basin. Elsewhere, there were open wharves for minerals including iron ore (*below*). (*Above: RCHS Slide Collection 75251. Below: Heartland Press Collection 867706*)

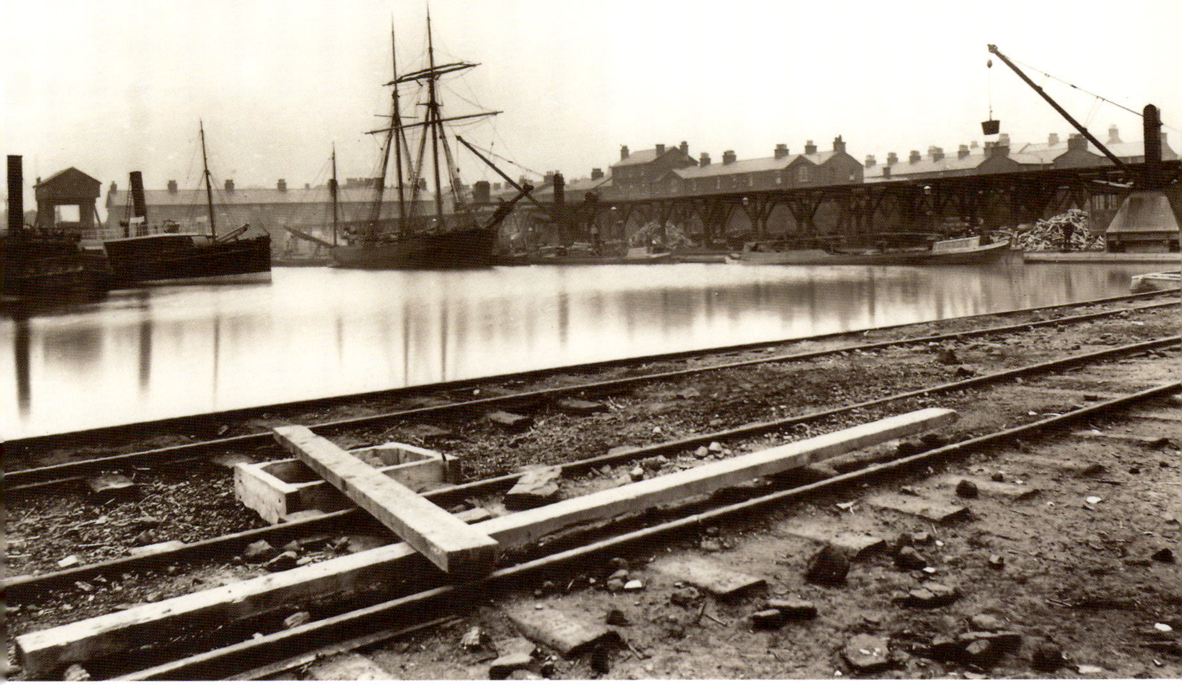

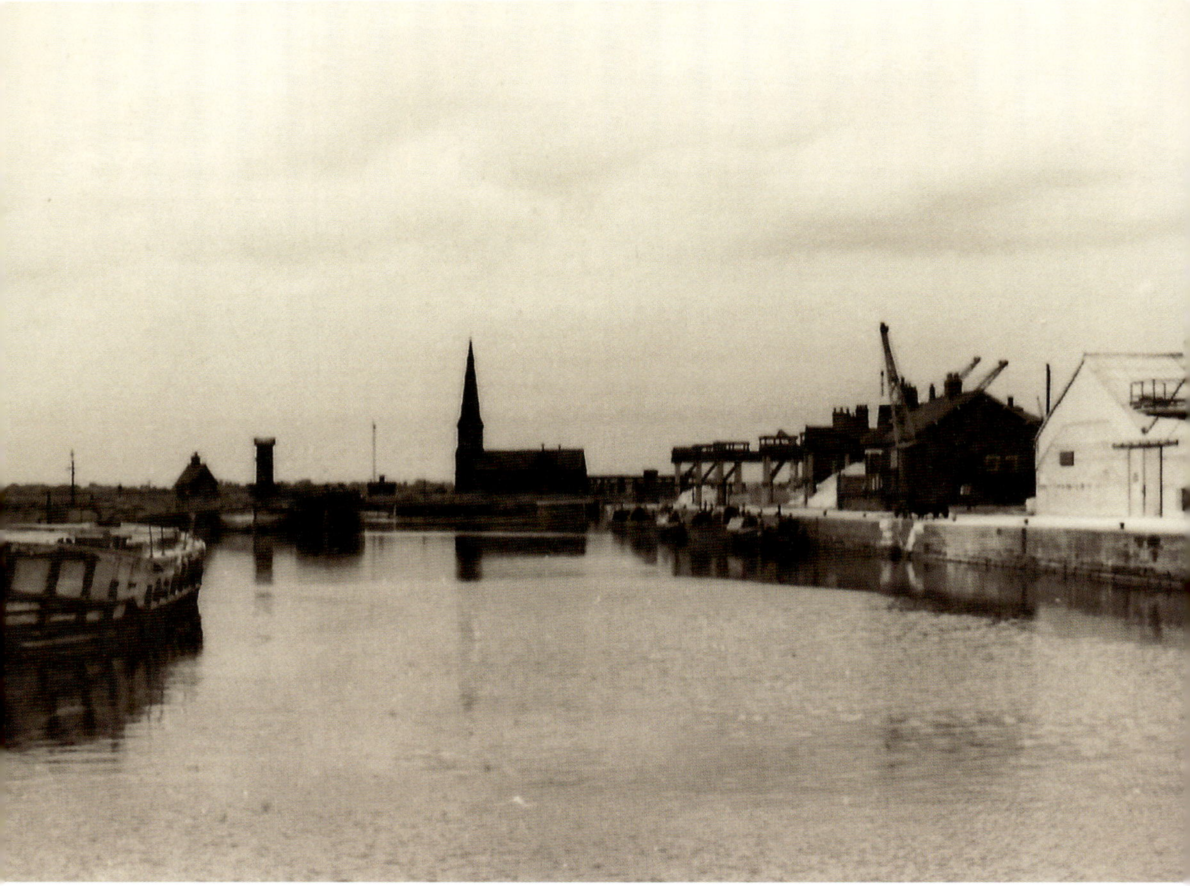

Delamere and Tollemache Docks, Weston Point, Weaver Navigation
The docks at Weston point terminated Delamere Dock, which originally had access to the Mersey. It was built during the mid-1860s to provide access for larger vessels to the end of Weston Canal. Another large dock, Tollemache, was constructed during 1885 on the east side of Delamere Dock, providing additional deep moorings between the Mersey and the Weston Canal. (*Railway & Canal Historical Society, H. Compton Collection, photographer P. Norton 65895*)

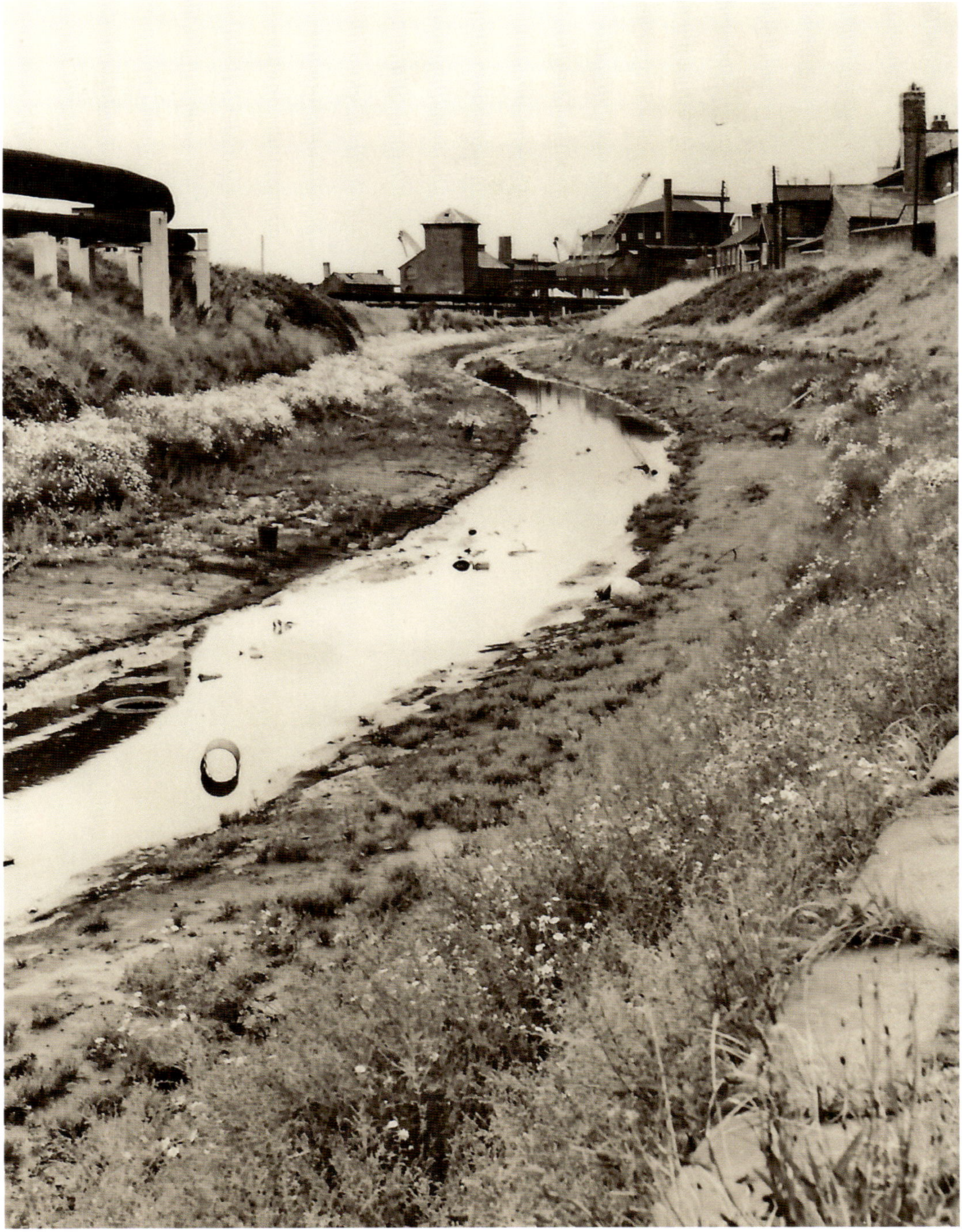

Runcorn, Runcorn & Weston Canal, 1957

The drained bed of the canal from Runcorn to Weston is shown here. This waterway saw little use after 1939, having never been a busy trading waterway which was the hope when this link was opened in 1859. Abandonment was not made official until 1966, which was the year the New Line of Runcorn Locks was abandoned. (*Railway & Canal Historical Society, H. Compton Collection, photographer P. Norton 65414*)

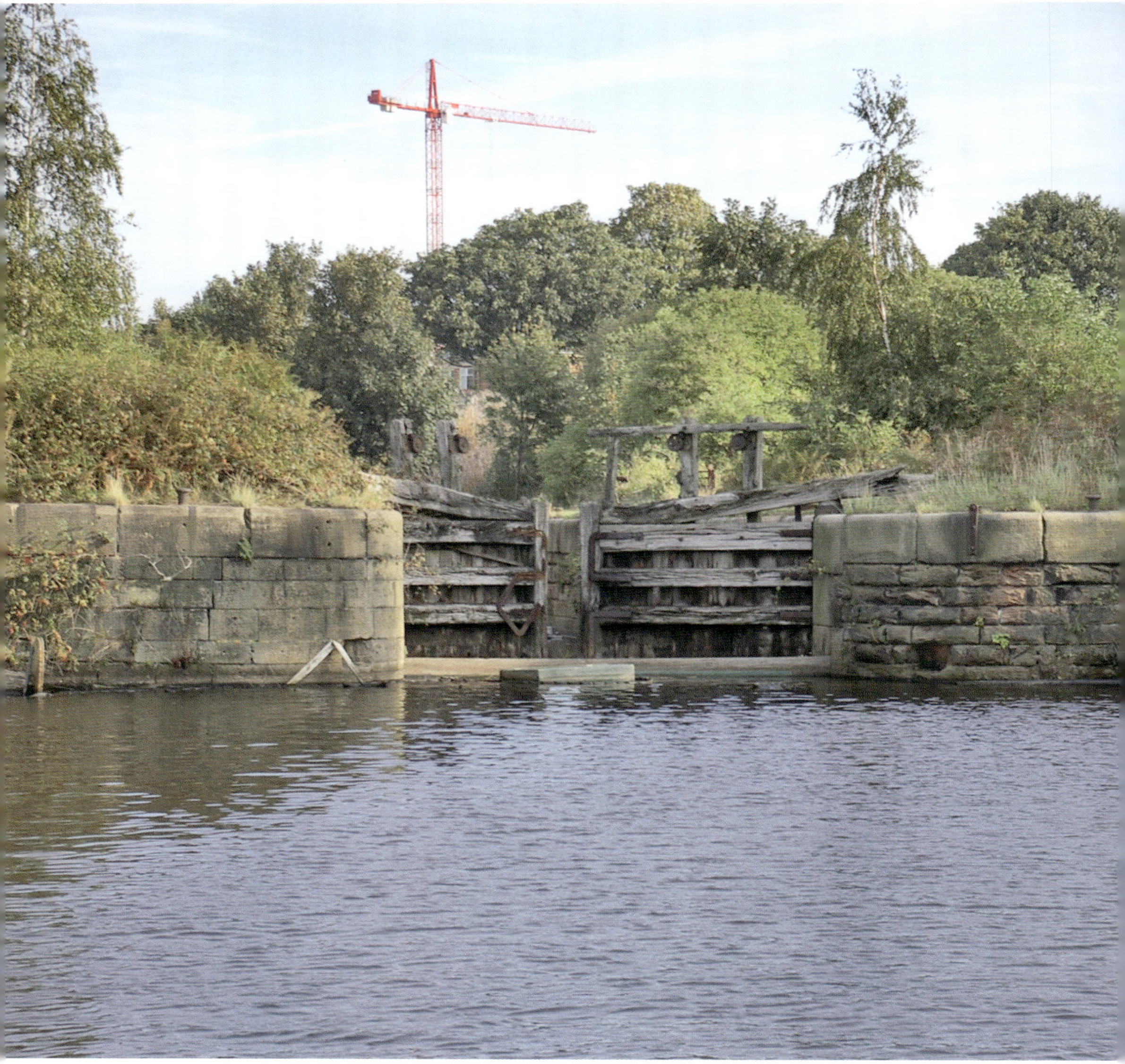

Runcorn & Weston Canal, Entrance Lock, Weaver Navigation
The entrance lock from the Weston Canal to the Runcorn & Weston Canal remains intact. This navigation was made for the Bridgewater Canal Company specifically for the transport of salt. (*Ray Shill*)

Chapter Four

Liverpool & Links to the Waterway Network

Leeds & Liverpool Canal, Port of Liverpool

Liverpool's dock network began east of the area known today as Mann Island, or Pier Head, and gradually extended both east and west where a collection of small and large docks were constructed. Some were tidal basins and others, whose entrances were protected by lock gates, had connections to others.

Making the first docks was entrusted to Thomas Steers, river navigation and canal builder, who became the first engineer. His place was taken by another canal engineer, Henry Berry, and between them developed the basic dock features. Later, other engineers improved on the early dock network as it expanded along the river front to the north and to the south.

Apart from coastal and seagoing vessels, Liverpool Docks were also served by the estuary sailing vessels, commonly known as flats. It was these flats that travelled onto some of the local canals and river navigations and provided the principal means of merchandise interchange with canal narrowboats at Ellesmere Port, Preston Brook and Runcorn.

The Leeds & Liverpool Canal initially terminated near the docks, but had no connection with them. Any trans-shipment between canal craft and river craft was made over land. A dock link was eventually made in 1848 when four new locks joined Stanley Docks with the canal main line.

Generally, canal-born goods came into the docks on flats that sailed up river to Runcorn or Ellesmere Port for the canal links to the Bridgewater and Ellesmere & Chester canals. Further upstream past Warrington such craft could either travel along the Sankey Brook Navigation to St Helens or the Mersey & Irwell Navigation to Manchester.

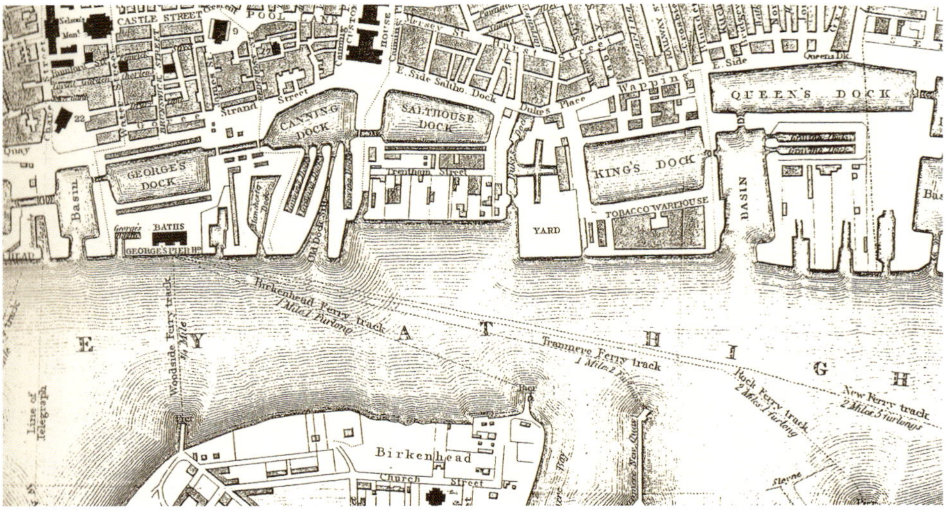

Liverpool Docks, Davis's Map of Liverpool, 1836

Liverpool Docks comprised different groups of tidal basins and inner basins isolated from the various tidal levels by gates. Merchandise carriers on wide and narrow canals had agents at George's, Duke's (centre), and King's or Queen's docks, to where traffic was consigned or despatched. Duke's Dock was often the destination of the boats that passed along Bridgewater Canal, and was, until 1900, the property of the Duke of Bridgewater and later his trustees. The Mersey & Irwell Navigation used Manchester Dock (centre). (*Heartland Press Collection 792011*)

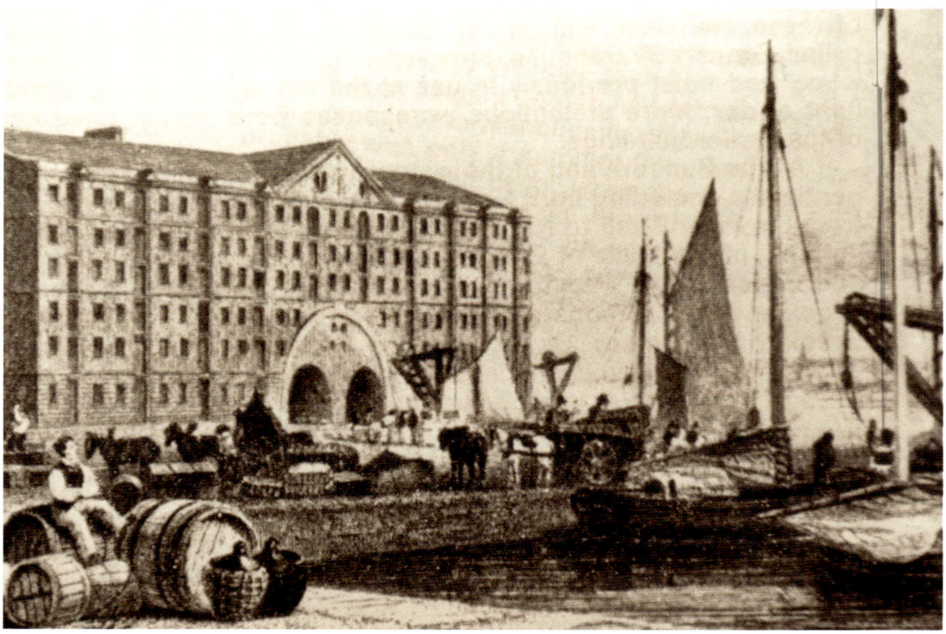

Duke's Docks, Liverpool

The Duke of Bridgewater's Dock at Liverpool became an important destination for flats carrying merchandise. The large warehouse was used to store corn. It was built in 1811. (*Heartland Press Collection*)

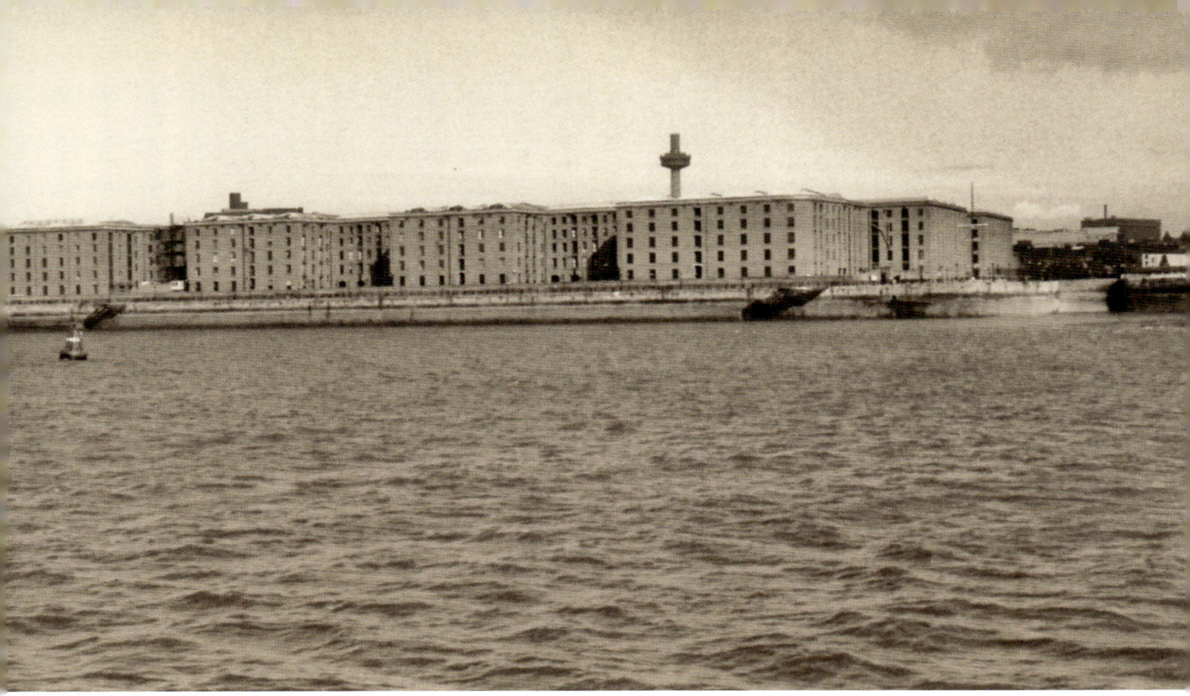

Liverpool Docks, Albert Dock

Albert Docks were opened by Prince Albert in 1846. They were built on the site of a former boatbuilding yard. They last saw use in 1972 and have since been restored for various mixed uses, including a maritime museum and art gallery. They are all Grade I listed. (*Ray Shill*)

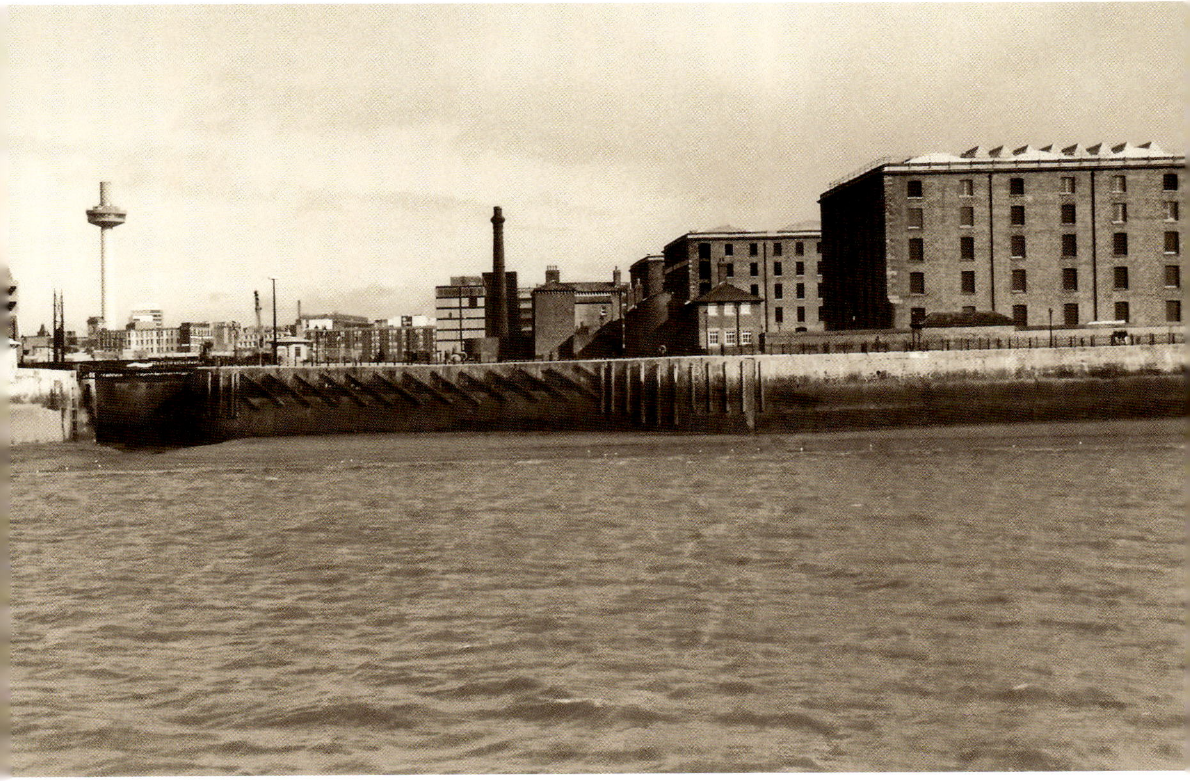

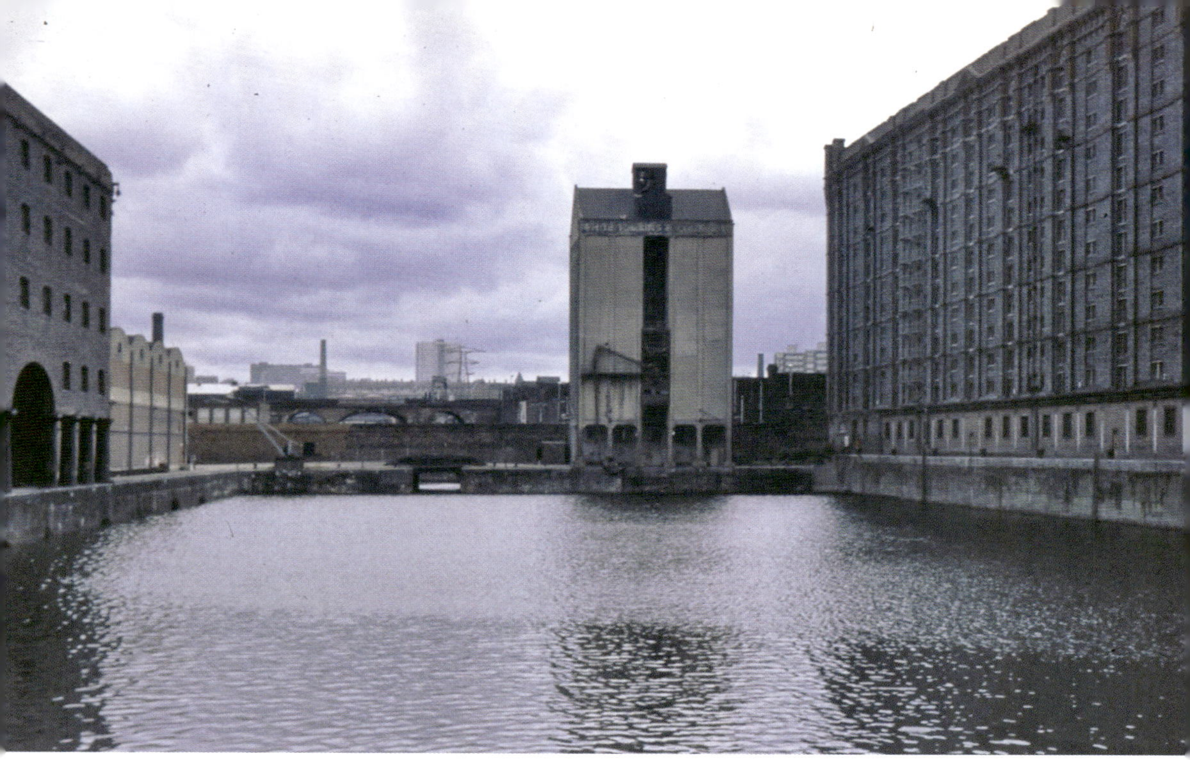

Stanley Dock, View of Dock (*Above*) & View of Jesse Hartley Warehouse (*Below*), Liverpool Docks, 1976

The Leeds & Liverpool Canal link entered the dock at centre of the upper image. The large building on the left was used to store tobacco. The lower image shows the Jesse Hartley warehouse, erected in 1848. (*RCHS Collection, Chris Bartlett, 82170 & 82171*)

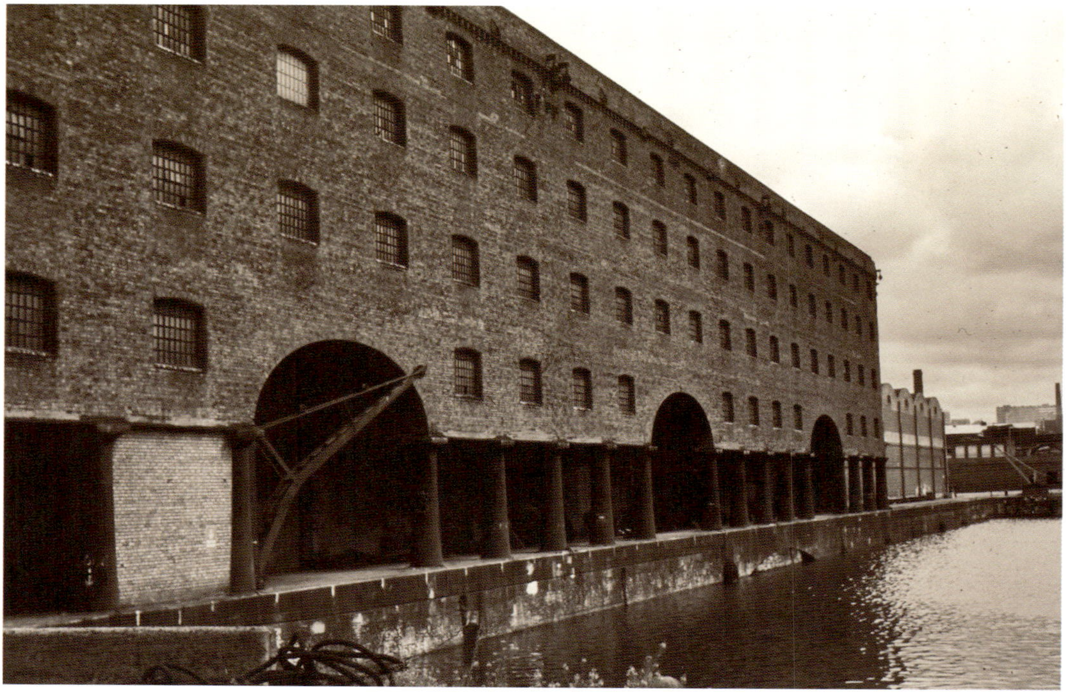

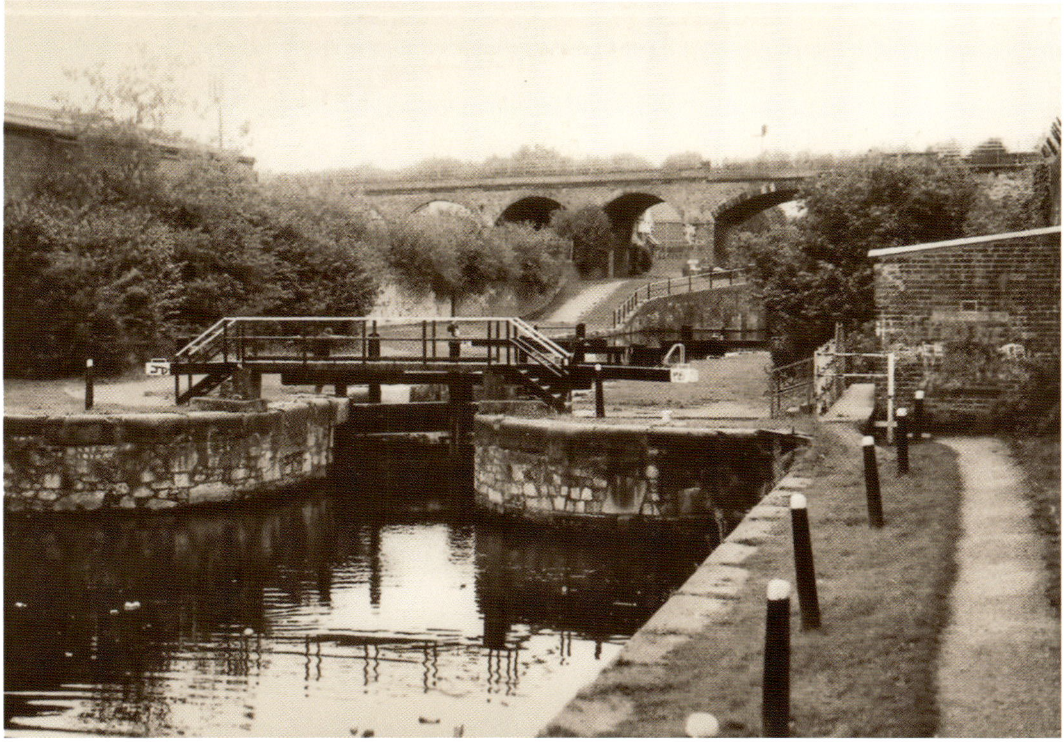

Stanley Lock Branch, Bottom Lock & Third Lock, Leeds & Liverpool Canal, 2007
The link to Stanley Docks from the Leeds and Liverpool main line was made in 1848. (*Ray Shill*)

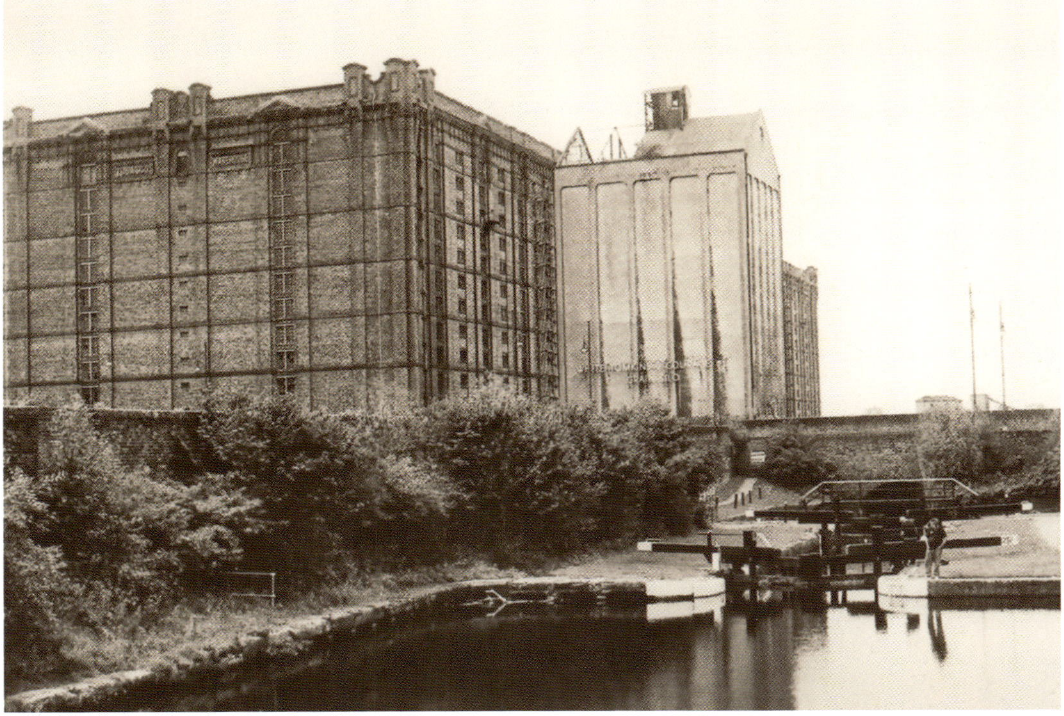

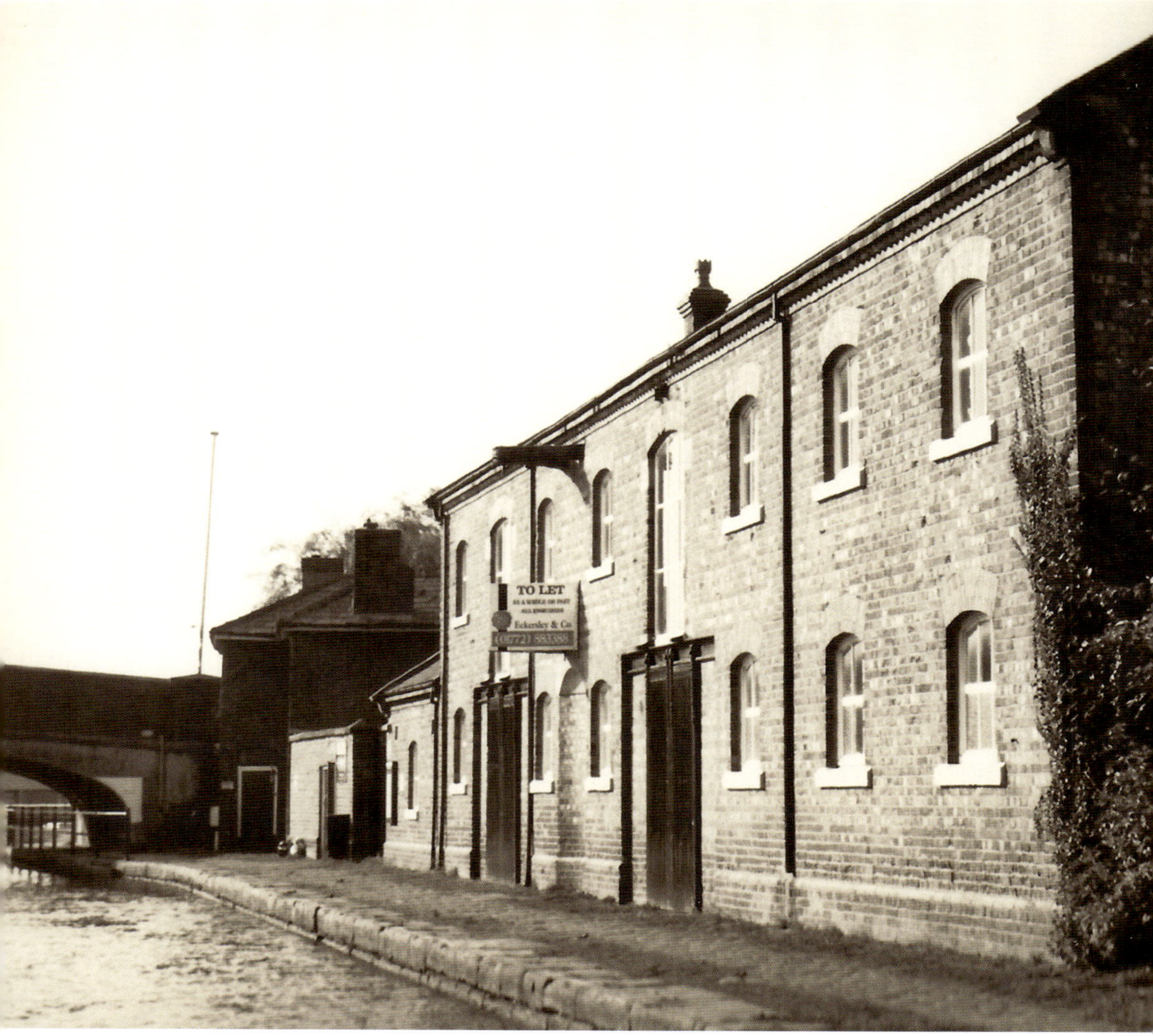

Burscough Warehouses, Liverpool, Leeds & Liverpool Canal
Survivors from the carrying age are the towpath warehouses and related buildings at Burscough. (*Ray Shill*)

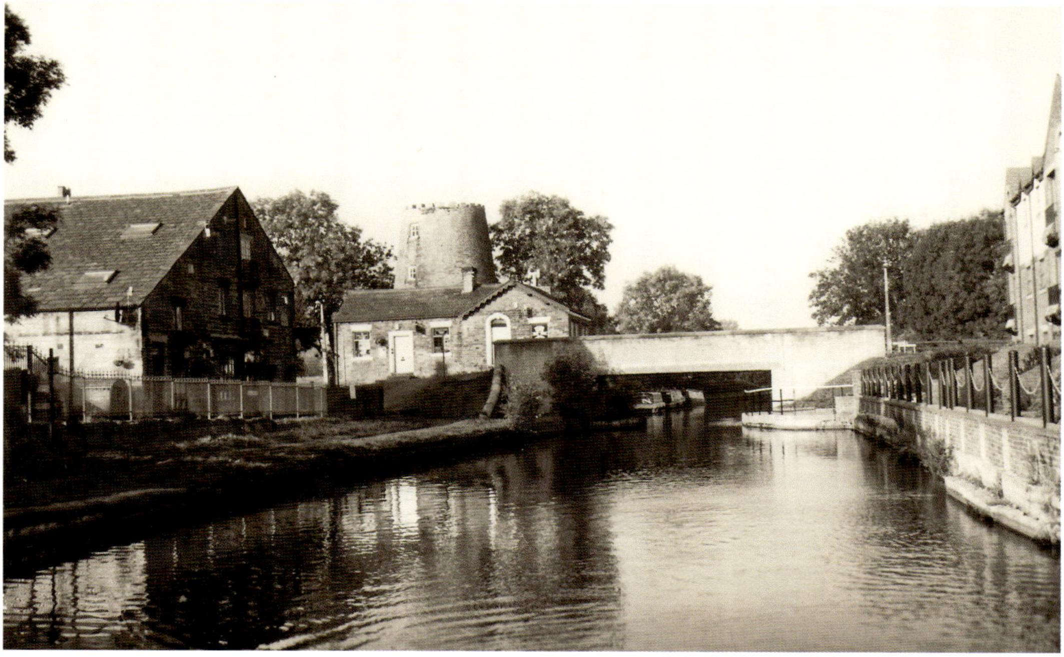

The Canal at Parbold, Leeds & Liverpool Canal

It was near Parbold that the Leeds & Liverpool Canal from Liverpool met up with the Douglas Navigation to Wigan. Also near Parbold, the canal crossed over the Douglas by an aqueduct that was high enough for craft on the Douglas to pass below. (*Above: Ray Shill. Below: Railway & Canal Historical Society, Post Card Collection 96492*)

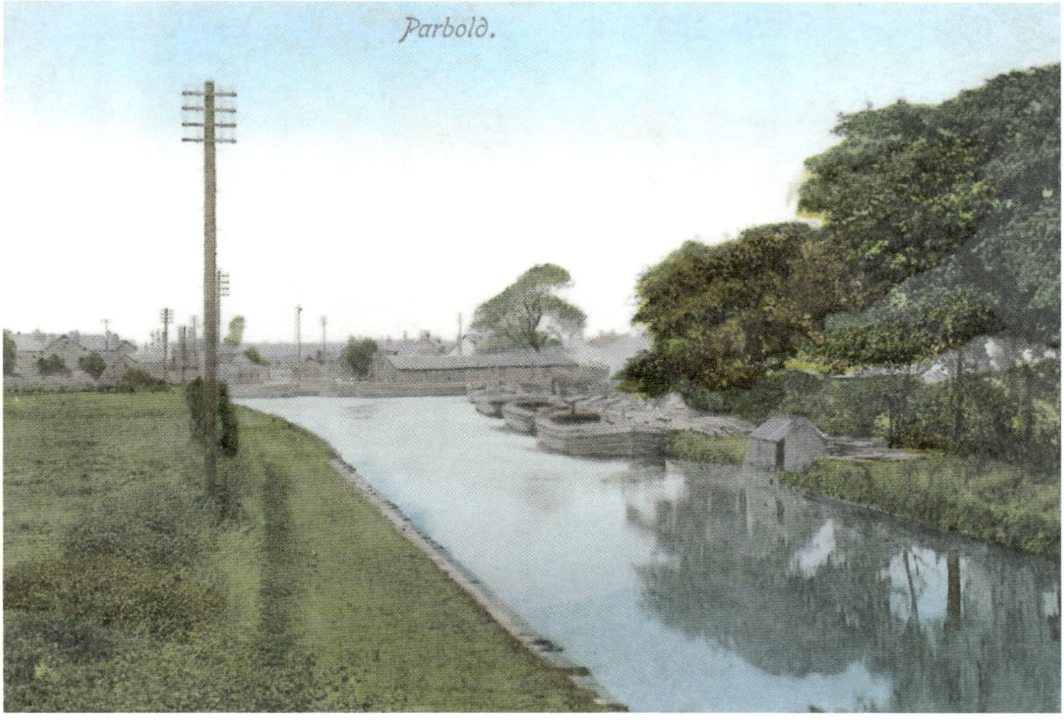

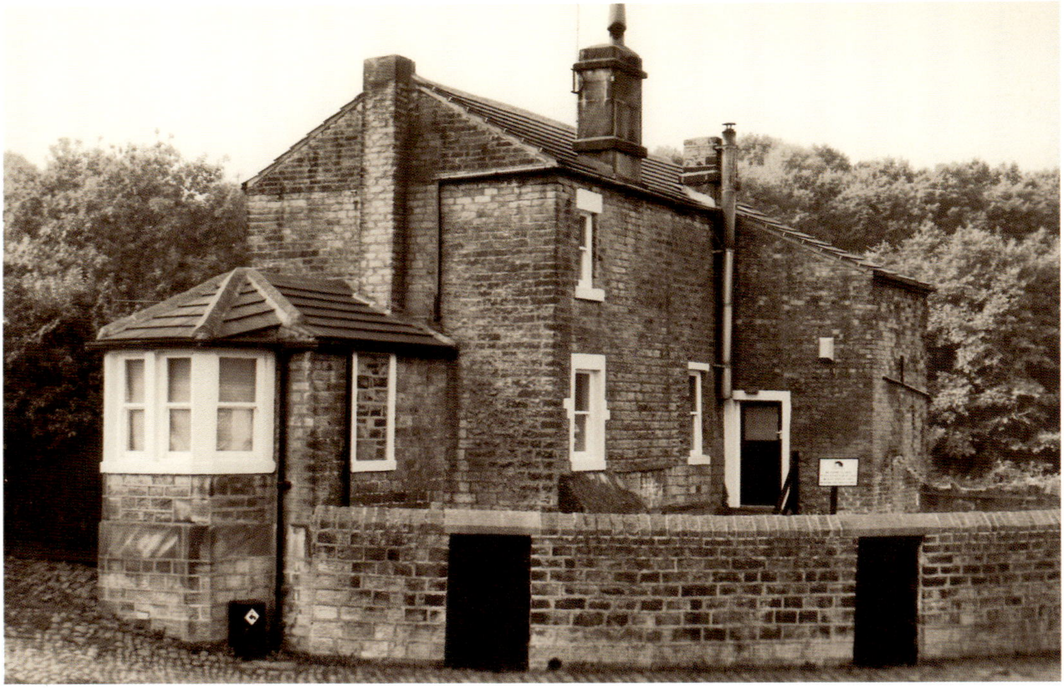

Lockhouse and Douglas River Lock, Dean, Liverpool, Leeds & Liverpool Canal
Extending the canal through to Wigan from Liverpool was done in stages, and was helped with the purchase of the Douglas Navigation. An extension of the new canal from Liverpool, known as Leighs Cut, built during 1772 was made as far as Dean, where a lock enabled craft to pass onto the Douglas River Navigation and the coal wharves there. (*Ray Shill*)

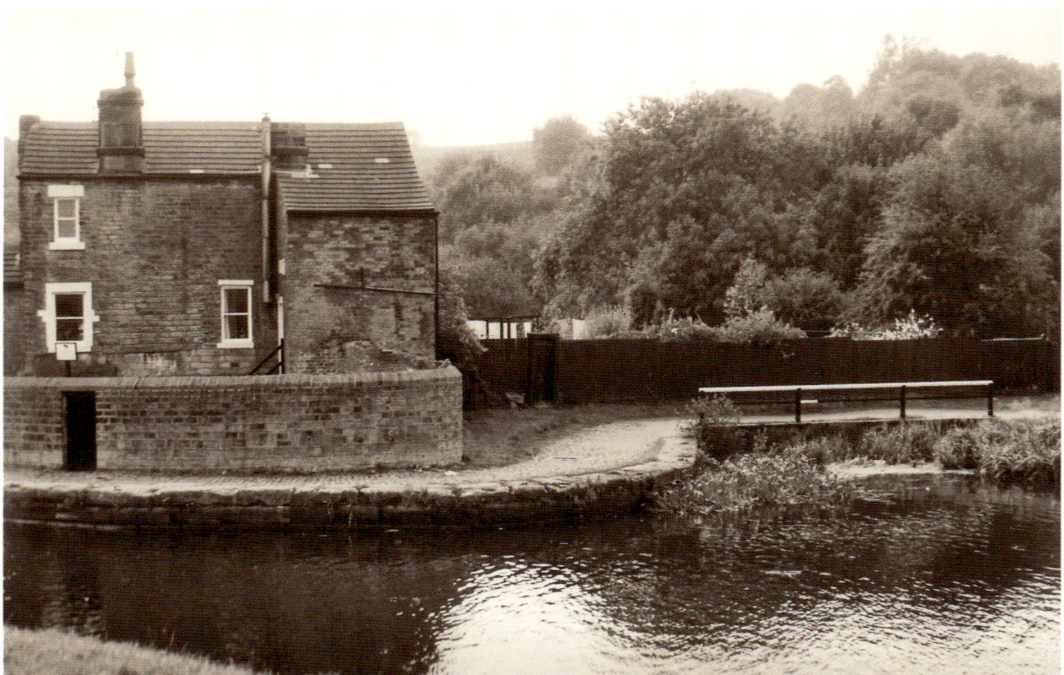

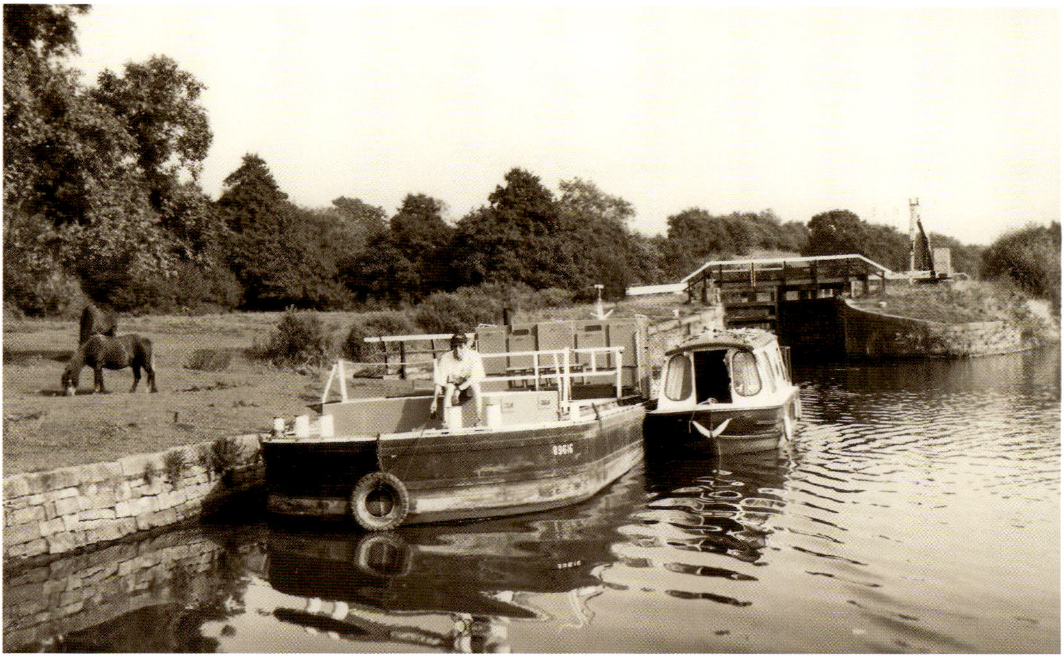

Appley Bridge Locks, Liverpool, Leeds & Liverpool Canal
The canal through Appley Bridge had a single deep lock, but during 1792 two locks were built to reduce water loss. The lock in the centre of the picture formed the lowest of the pair built then. (*Ray Shill*)

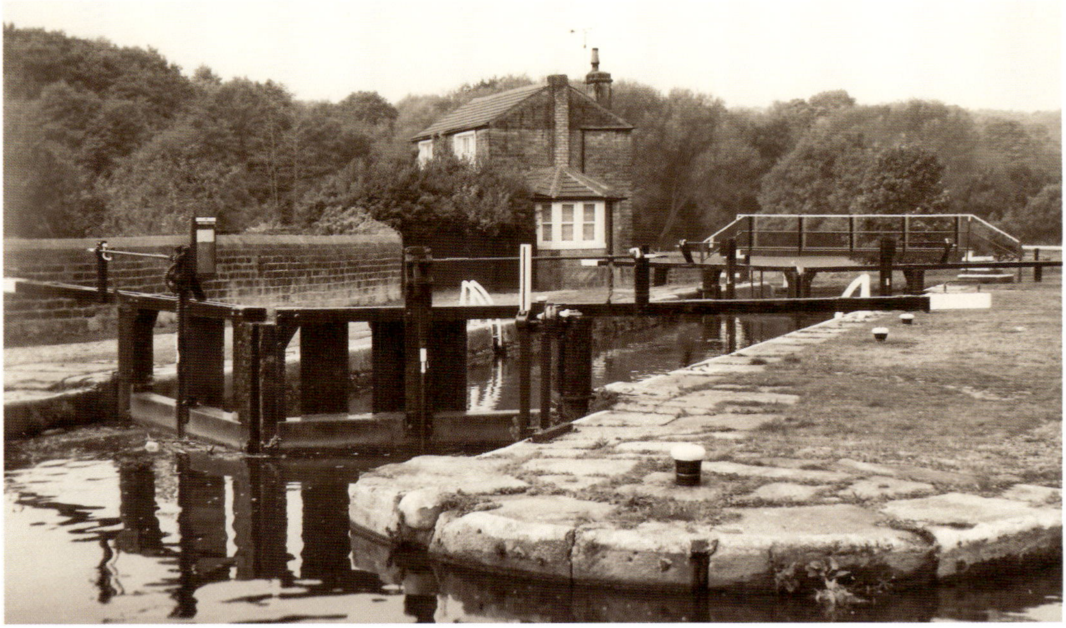

Dean Locks, Liverpool, Leeds & Liverpool Canal
The main canal was extended from Dean to Wigan, which included the making of Dean Lock in 1781. Increasing trade led to the making of the adjacent lock later. (*Ray Shill*)

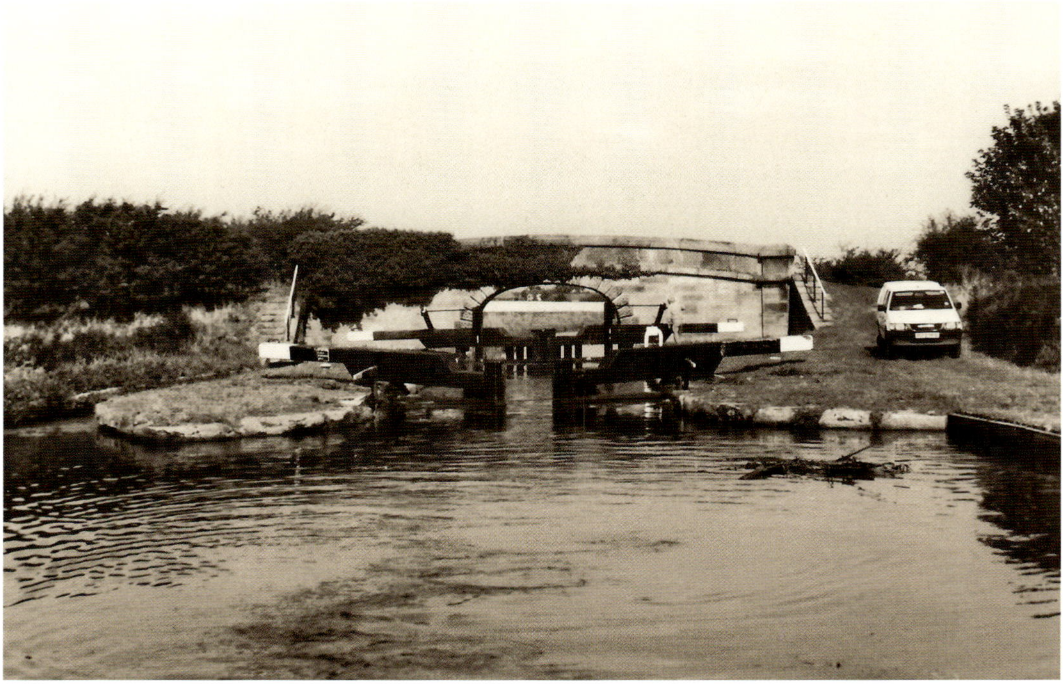

German's Lock (*Above*) & Tarleton Lock (*Below*), Rufford Branch, Leeds & Liverpool Canal
The Rufford Branch provided a link between the Leeds & Liverpool Canal and the Douglas Navigation, initially to Sollom and then Tarleton. At Tarleton, craft were able to lock down into the navigable River Douglas in order to reach the Ribble and Preston. (*Ray Shill*)

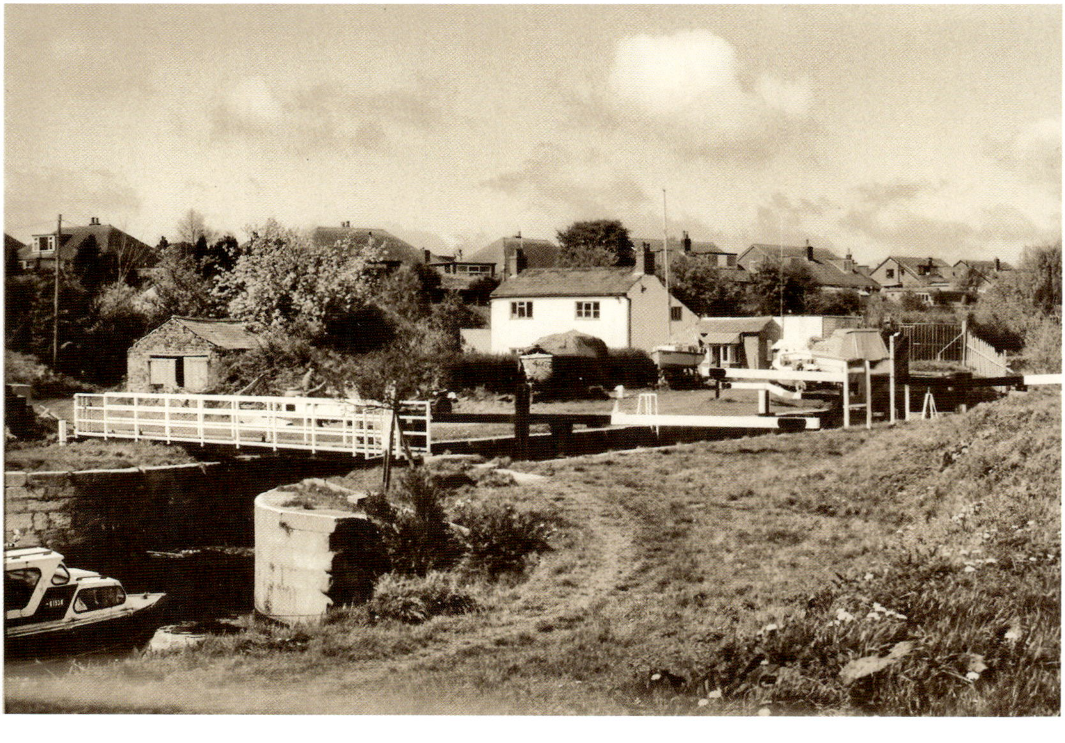

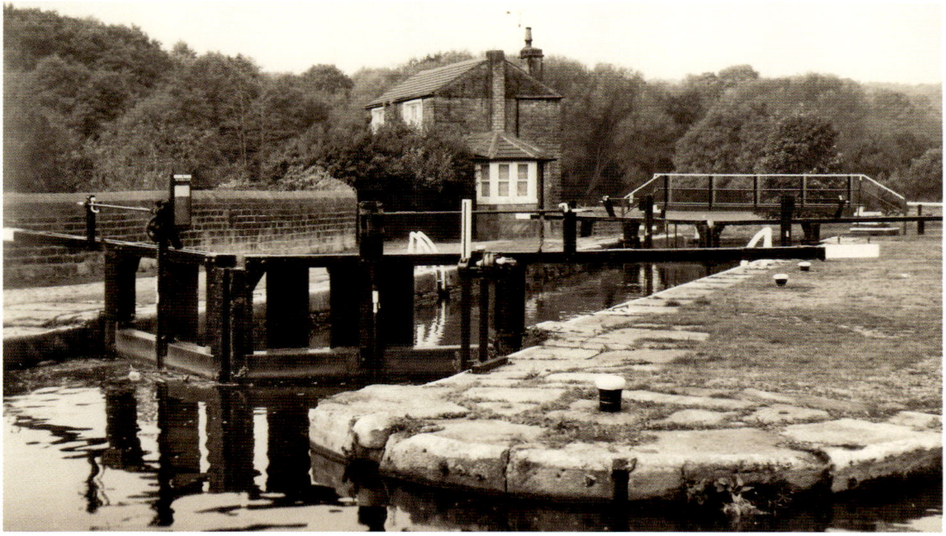

Warehouses, Wigan, Leeds & Liverpool Canal

There were a group of warehouses at Wigan placed at a right-angled turn in the waterway. These formed the main carrier's establishment for the town. There were five separate warehouses erected beside the canal; the first were put up when the canal reached this spot and the others erected at different times throughout the nineteenth century. From here the main canal turned right and climbed up Wigan Locks to Kirklees. This area was known as Wigan Pier, as the Winstanley Colliery railway terminated here and loaded boats from the right-hand bank. (*Ray Shill*)

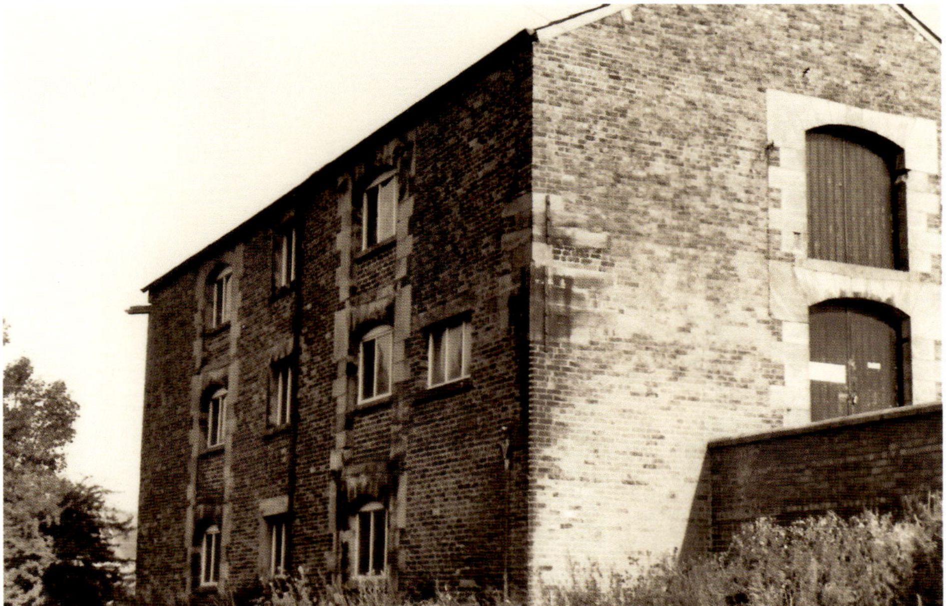

Warehouse (of 1803) Near Tarleton, Rufford Branch, Leeds & Liverpool Canal

The warehouse at Tarleton was put up at the time of the canal and river improvements made through the Croston Drainage scheme. (*Ray Shill*)

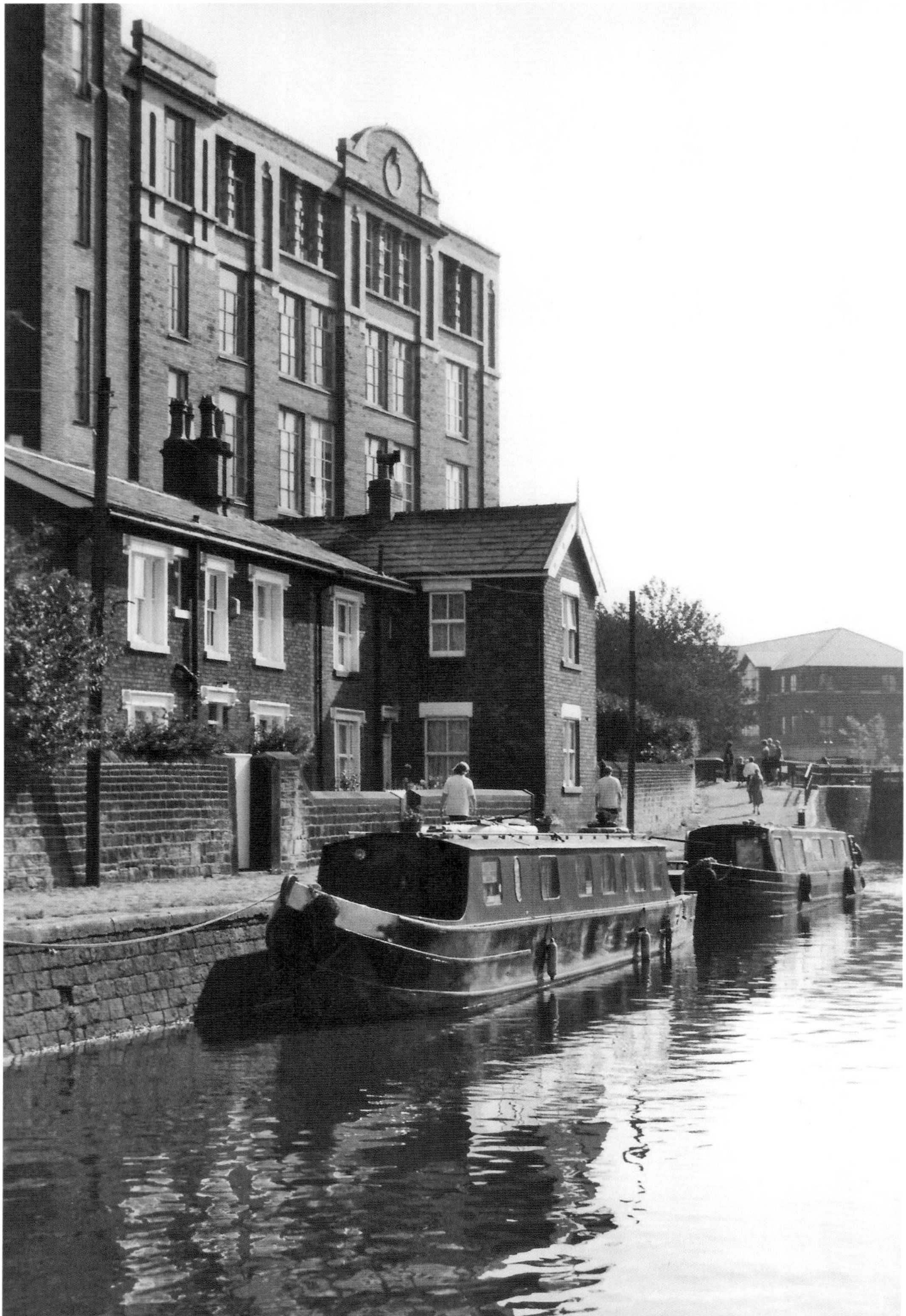

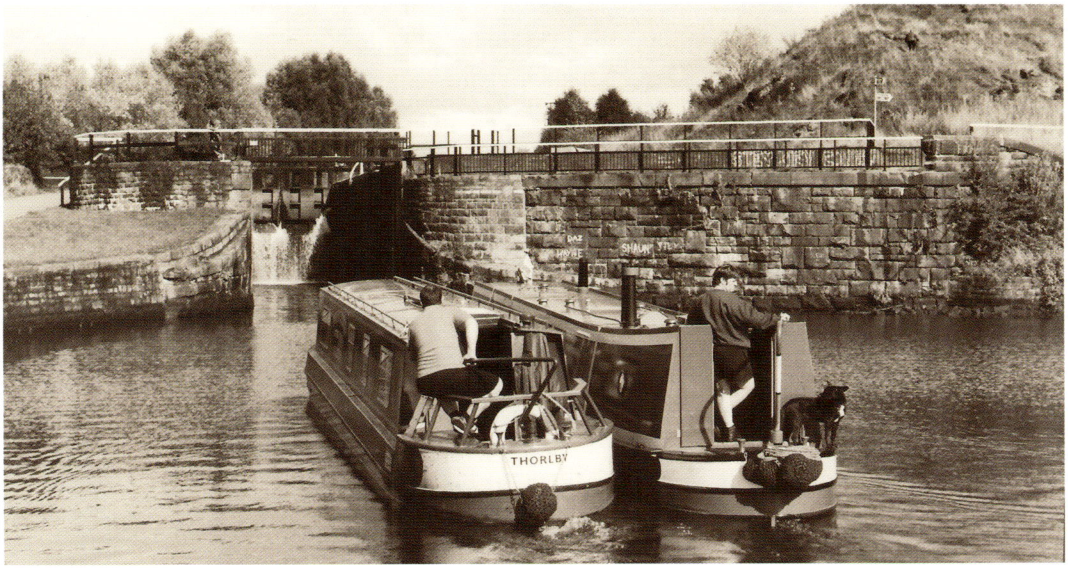

Wigan Locks (*Above*) and Kirklees Ironworks (*Below*), Leeds & Liverpool Canal
The Wigan Coal and Iron Company had, during the nineteenth century, an extensive mining and iron making industry. Iron was smelted at Kirklees Ironworks, which was placed beside the canal. (*Above: Ray Shill. Below: Heartland Press Collection 768166*)

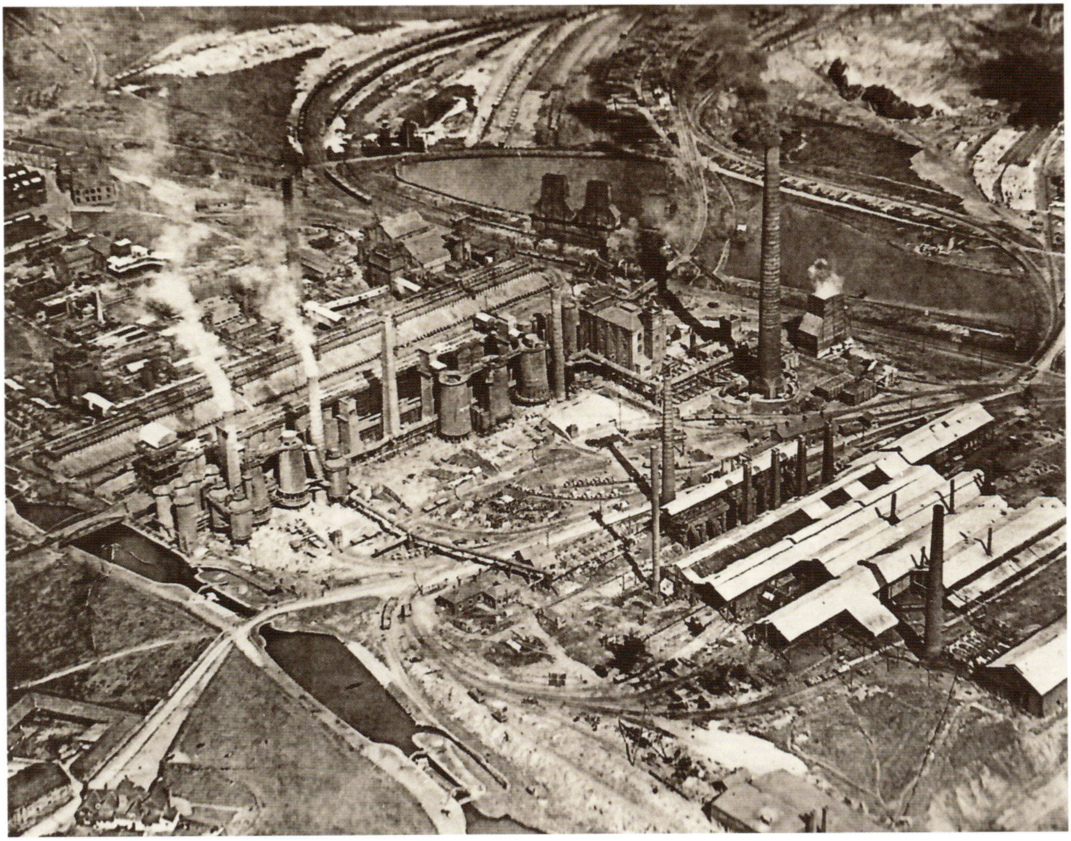

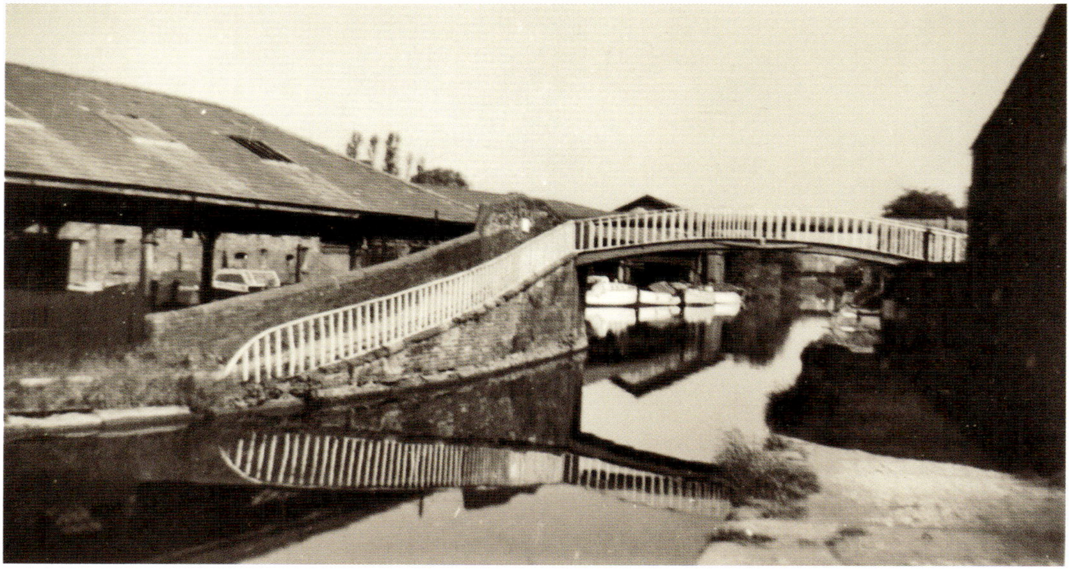

Iron Turnover Bridge, Tower Wharf & Junction with Dee Branch, Ellesmere Canal
The first section of the Ellesmere Canal opened from Tower Wharf to Whitby in 1795. That route joined the Chester Canal at a point placed above the original second lock and was accompanied by the making of two new locks to the tidal basin that joined the River Dee. Tower Wharf was placed on the far right of this image, while the iron turnover bridge was provided to convey the towing path over the waterway. (*Railway & Canal Historical Society, Hugh Compton Collection 70158*)

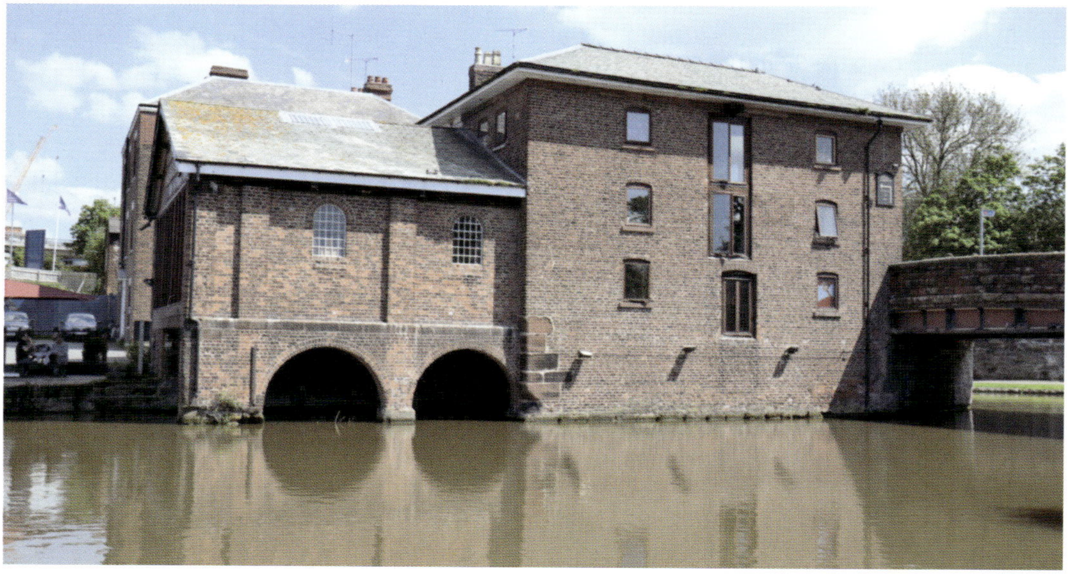

Tower Wharf, Warehouse & Junction with Chester Canal, Ellesmere Canal
Tower Wharf was a major commercial depot for the Ellesmere Canal when it first opened in 1795 and continued in this role for their successors, the Ellesmere & Chester Canal and the Shropshire Union, which have been restored and are retained as a feature in the public house that now occupies the structure. The large building behind was the offices for the canal company. (*Railway & Canal Historical Society, Hugh Compton Collection 75168*)

Chapter Five

Railway Features Along the Waterway

Railways and tramways linked with both the Leeds & Liverpool Canal and Douglas Navigation in the colliery area around Wigan. Tramways also linked mines at St Helens with branches of the Sankey Canal. There was also a stone block railway that linked Storeton Quarries with Bromborough Pool.

The first public railway in this area was the Liverpool & Manchester that was opened in 1830 and crossed the Sankey Canal by a viaduct. Other railway schemes followed, including the Wigan Branch Railway which joined the Liverpool & Manchester and crossed the Leeds & Liverpool Canal south of this town. A principally freight only line, complete with inclines, was made from the collieries at St Helens to a new dock on the north bank of the Mersey, in an area known as Runcorn Gap, but was renamed Widnes. The railway (St Helens & Runcorn Gap) was opened in 1833, in the same year as the Sankey Canal was extended from Fiddler's Ferry to a new lock and dock on the west side and adjacent to the railway dock.

Other railways followed, crossing navigations often by imposing structures. The longest viaduct was at Dutton, which passed over the Weaver floodplain and carried the Grand Junction Railway over the river here. There was also another crossing of the Weaver near Hartford (Vale Royal) and both the Mersey and Latchford & Runcorn Canal, south of Warrington.

Between 1844 and 1860, there was a period of intense building of railway links across the country, and many more lines were built with viaducts over waterways. In this part of the North West, railway construction continued for a few years later. The Cheshire Lines Committee had brick and iron viaducts at Northwich and Sankey. The London & North Western Railway extensions included the line through Runcorn and the long viaduct and bridge over the Mersey and Manchester Ship Canal.

Railway interchange facilities were made at Saltney beside the Dee where coastal vessels could interchange minerals and goods with the Great Western Railway. There was an important traffic in moving Furness ironstone by sea and rail to South Staffordshire furnaces at one time. Later, interchange depots were made with the London & North Western Railway at Preston Brook (Trent & Mersey Canal) and Calverley (Shropshire Union).

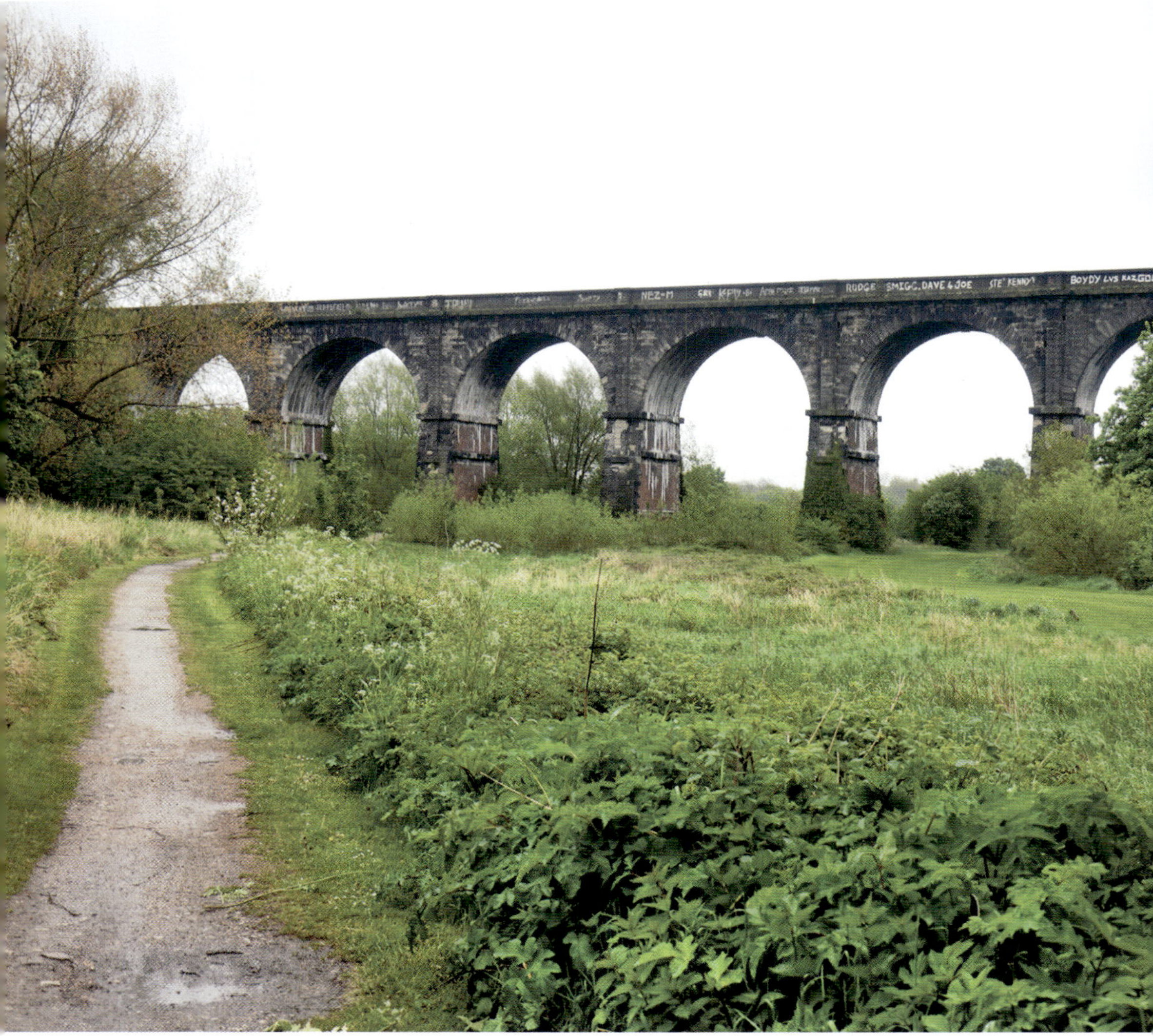

Sankey Viaduct, Sankey Brook Navigation

Railway and tram road viaduct construction entered a new age with the making of the nine-arch Sankey Viaduct across the valley of the Sankey Brook and the adjacent canal. This viaduct carried the Liverpool & Manchester Railway across the valley and marked an important milestone in transport history. It was built in 1830 in brick, and was faced with stone supplied by Thomas Brassey from Storeton Quarry in Cheshire. (*Ray Shill*)

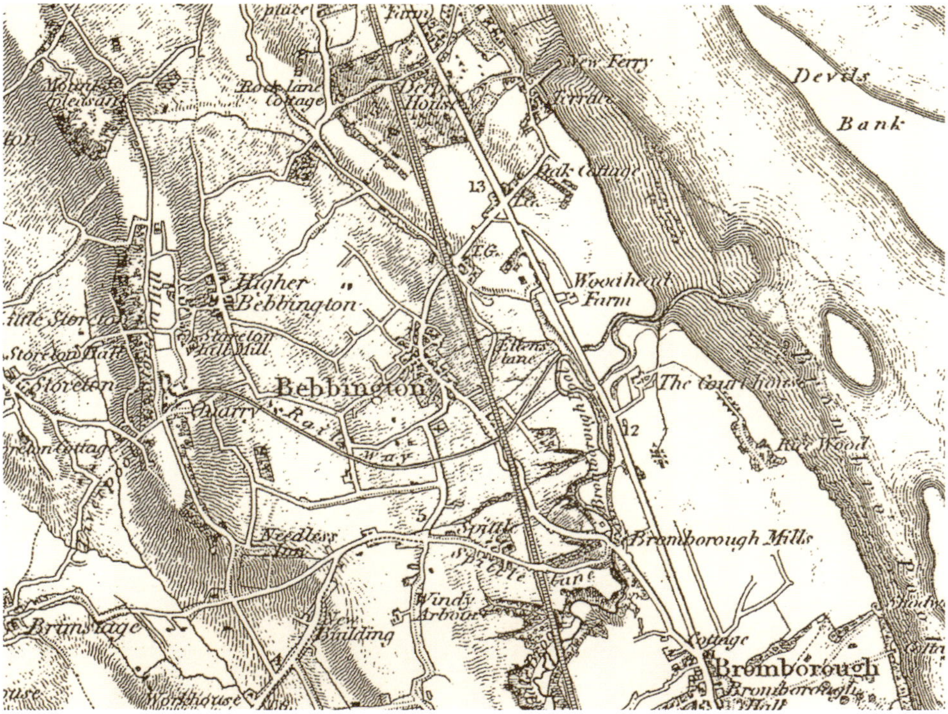

Storeton Railway Route, Port Sunlight

The route of the Storeton Railway ran from quarries near Bebbington to a pier alongside Bromborough Pool. It was an edge railway laid on stone blocks, opened in 1838. Lever Brothers Port Sunlight works were developed on adjacent land. In the lower image, the route of the railway passed under the railway and to the left of the Soap Works. (*Above: Ordnance Survey (793215). Below: Railway & Canal Historical Society, Bertram Baxter Collection 22439*)

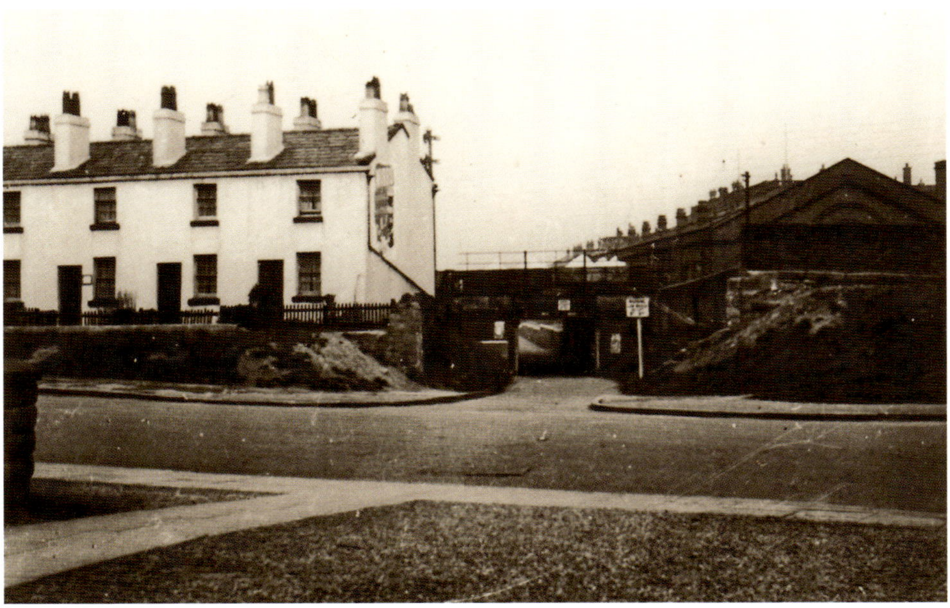

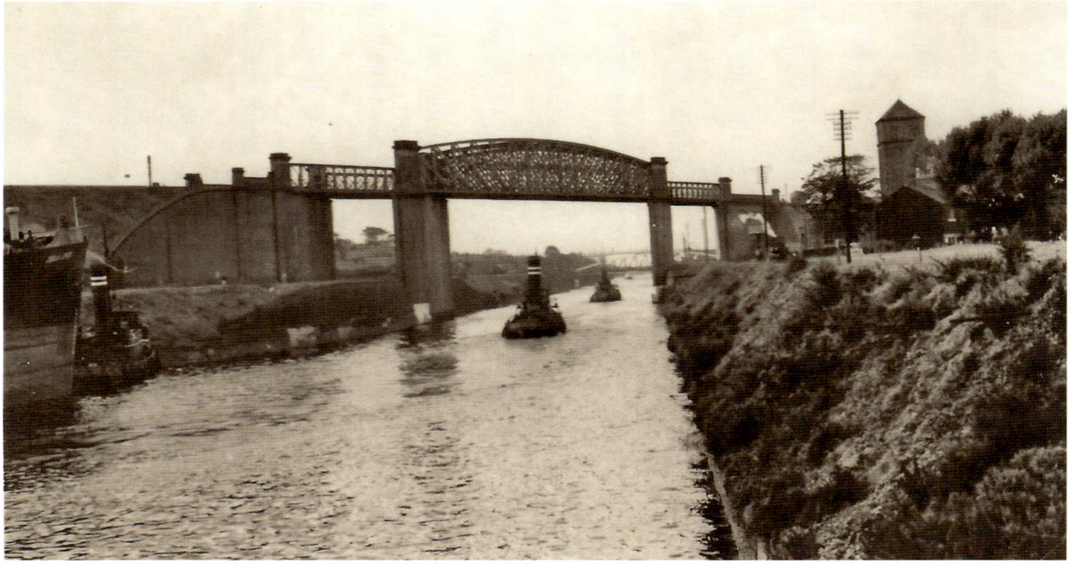

Latchford, Manchester Ship Canal
The now disused Latchford Viaduct was completed in 1893 to carry the London & North Western Railway line from Stockport to Warrington over the Manchester Ship Canal. (*Railway & Canal Historical Society, P. Norton Ref No. 65153*)

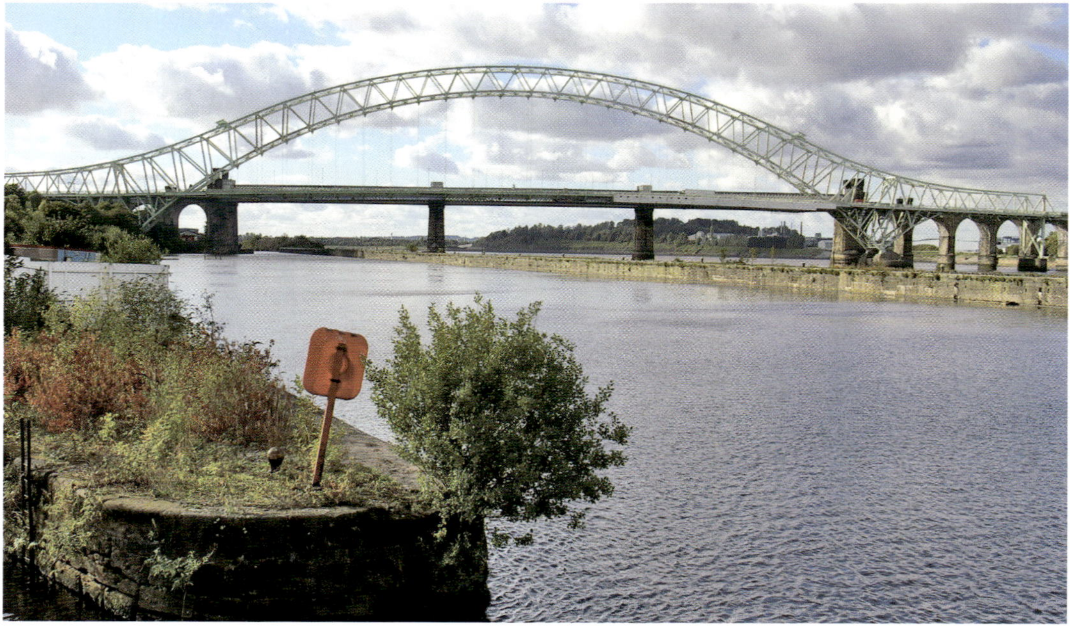

Road and Railway Bridges at Runcorn, River Mersey and Manchester Ship Canal
The distinctive road bridge in the foreground, opened in 1961, replaced a transporter bridge on this site. It links Runcorn with Widnes and is now known as the Silver Jubilee Bridge. The railway bridge behind it was built for the London & North Western Railway to the design of William Baker. Construction work was carried out between 1864 and 1868, but the bridge did not open for traffic until the next year. (*Ray Shill*)

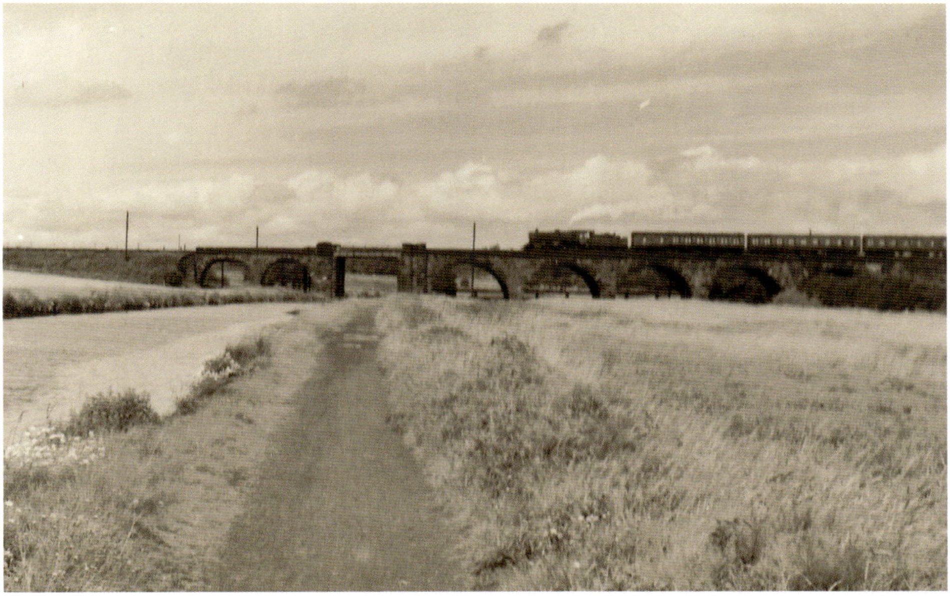

Cheshire Lines Committee, Railway Viaduct, Sankey Canal

The Cheshire Lines Committee route opened in 1873, between Manchester and Liverpool, and crossed the Sankey Canal near Warrington. (*Railway & Canal Historical Society, Hugh Compton Collection, P. Norton 65430*)

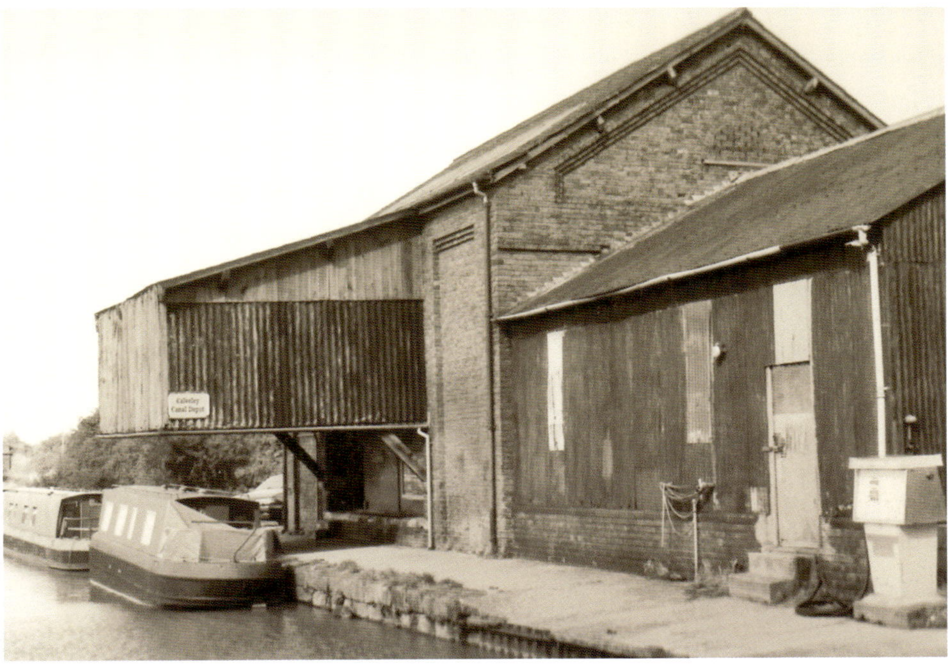

Railway Interchange Wharf, Calverley, Shropshire Union Canal

The railway between Crewe and Chester passed close to the Chester Canal at Calverley. Railway sidings and an interchange depot was provided here for trade passing between canal and railway. (*Ray Shill 864685*)

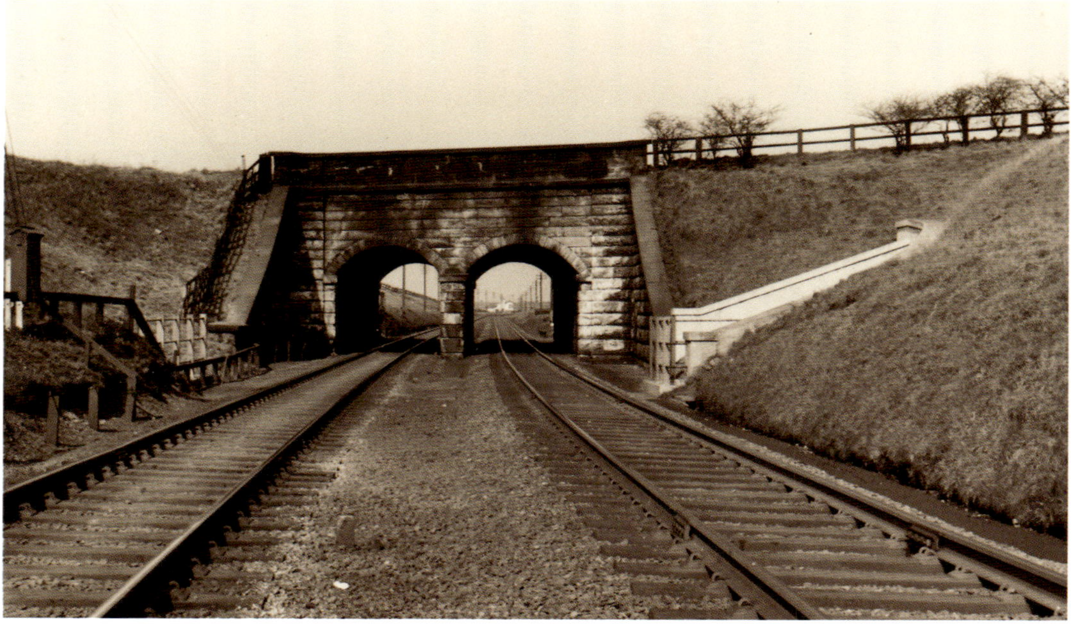

Bridgewater Canal Aqueduct, Preston Brook
The Grand Junction Railway passed under canal line to Runcorn through two arches. (*Railway & Canal Historical Society, Hugh Compton Collection, P. Norton 64285*)

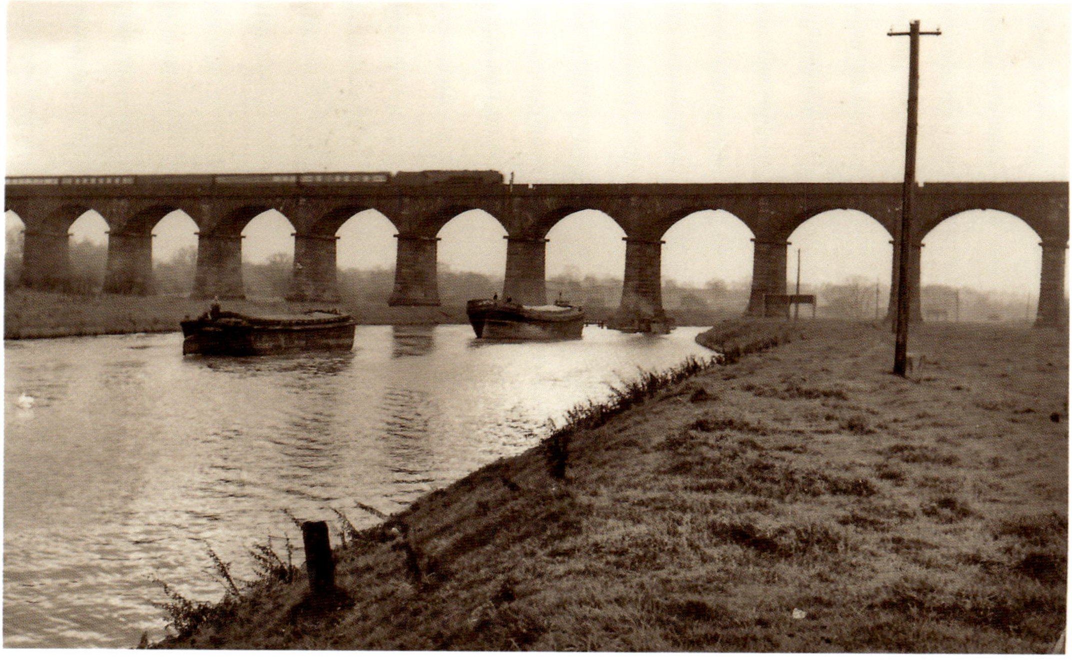

Dutton Viaduct, River Weaver
The massive viaduct (500 yards long) near Dutton was built to carry the Grand Junction Railway over the Weaver. Constructed in 1836 by the contractor William McKenzie, this viaduct comprises twenty-two arches. (*RCHS Hugh Compton Collection 65912*)

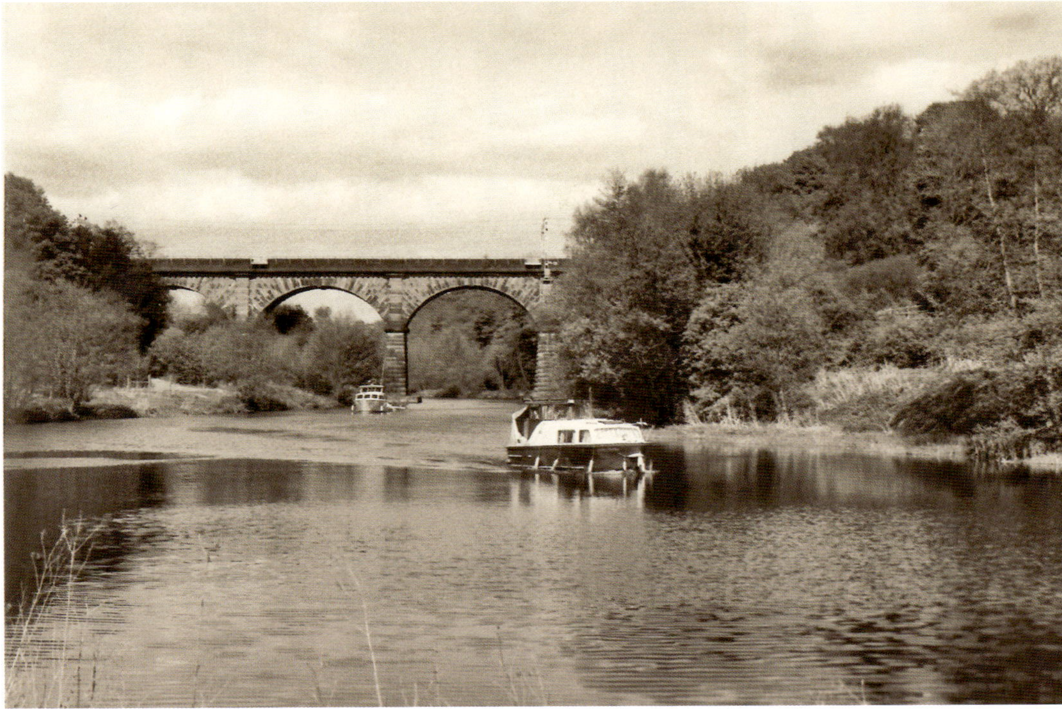

Vale Royal Viaduct, River Weaver

The shorter Vale Royal Viaduct crossed the Weaver near Hartford was completed by 1837 and has five arches, but only two span the river. (*Ray Shill*)

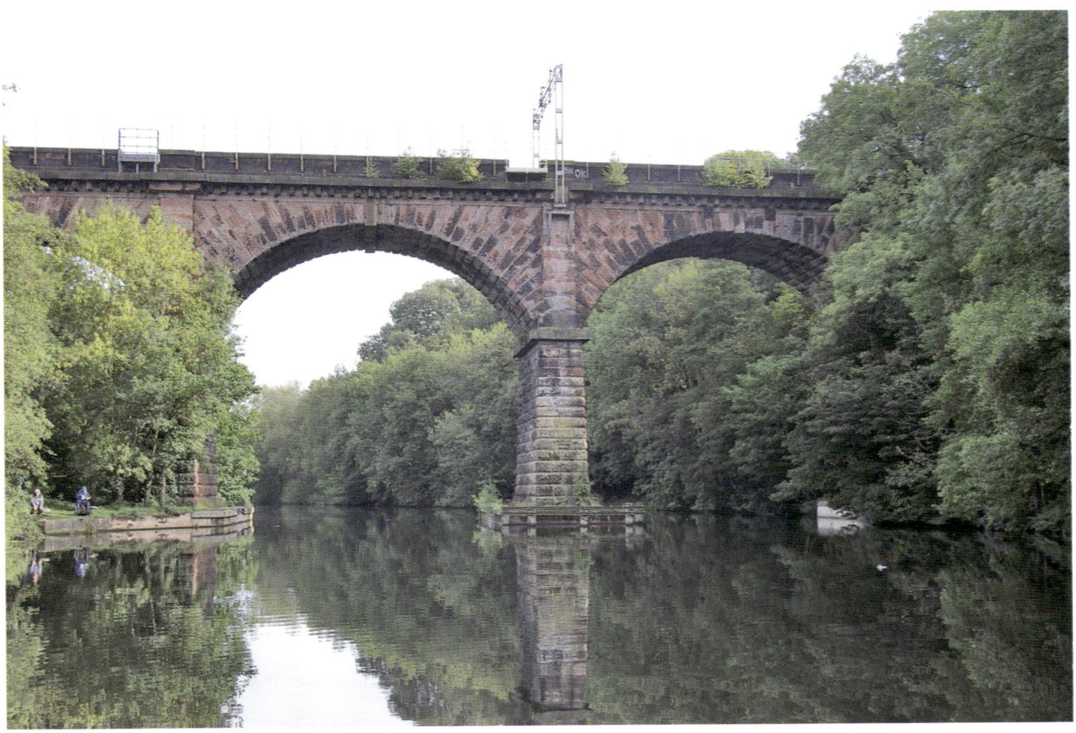

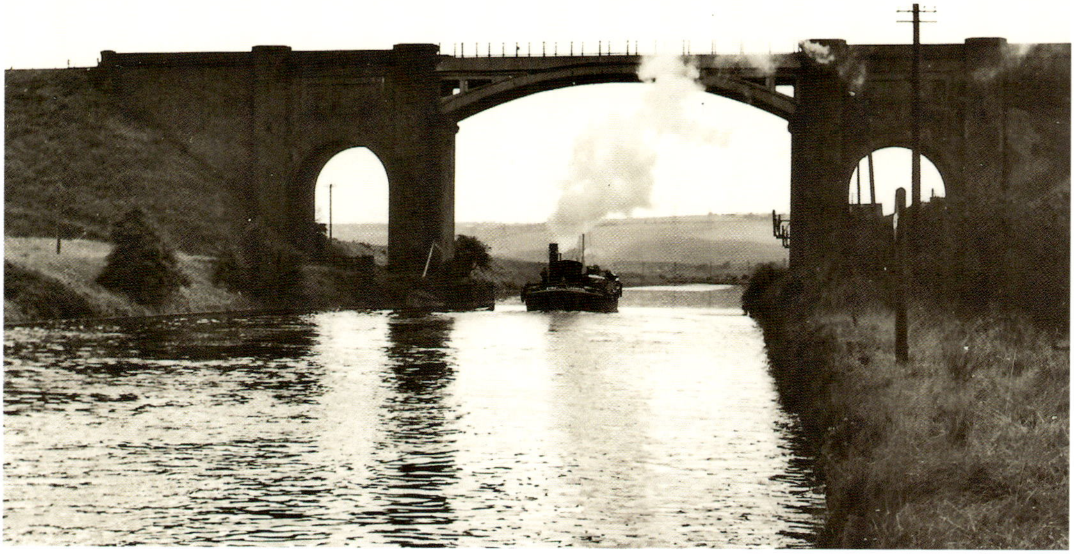

Sutton, Frodsham, Weston Canal

The brick, stone and iron viaduct was made to carry the Birkenhead, Lancashire & Cheshire Junction Joint Railway (Great Western & London & North Western Railway) over the Weston Canal in 1850. (*RCHS Hugh Compton Collection 65896*)

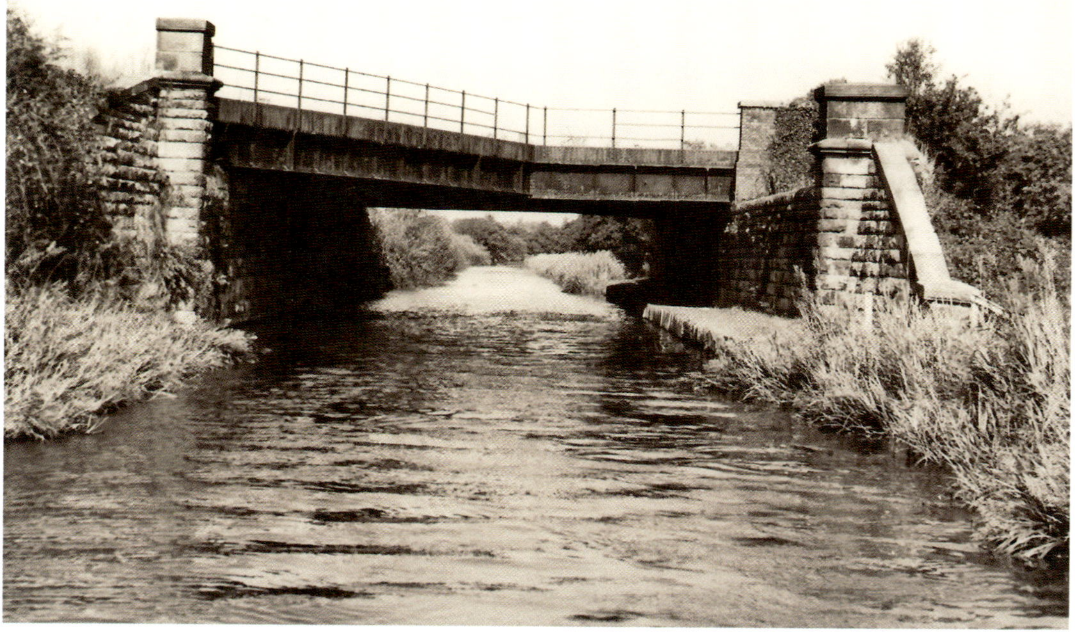

Liverpool & Preston Railway Bridge, Near Rufford

The Liverpool, Ormskirk & Preston Railway crossed the Rufford branch diagonally. (*Ray Shill*)

Chapter Six

Canal & River Improvements in the Nineteenth & Twentieth Centuries

Litherland Lift Bridge, Manchester Ship Canal, Shropshire Union Canal, River Weaver (Bridges and Anderton Lift)

The Weaver Navigation was made possible through the construction of locks that enabled flats to reach the town of Winsford. These locks were improved and enlarged at different periods. In fact, there was almost a continuous period of change from 1843–97. A major step was the provision of additional and larger lock chambers adjacent to the existing locks to create double locks, in a scheme proposed by engineer, William Cubbit. When this was finished, a further and major improvement scheme was instituted by Edward Leader Williams, engineer to the Weaver Navigation. The basic concept was the reduction of locks and different levels, building new weirs and bigger and deeper locks employing new technology in the movement of water into and out of the lock chamber. Edward had assisted his father with similar works on the River Severn, and now translated these concepts for the Weaver. He was also responsible for designing the Anderton Lift, which used hydraulic power to raise and lower boats between the Weaver and the Trent & Mersey Canal.

Steam towage on the river commenced in 1863, and this innovation encouraged further improvement. Between 1865 and 1897, the task of making the river navigable for seagoing vessels and reducing the mileage from Weston Point to Winsford was carried out under a team of four engineers. The navigation was shortened by 10 miles, levels between locks were raised and water held back by massive sluices. The number of locks on the main river section was reduced to four (Dutton, Saltersford, Hunts and Vale Royal). Leader Williams went onto to be engineer for the Bridgewater Trustees, leaving others to take his place as engineer to the Weaver Navigation and carry out and amend the project, which eventually saw the reduction of locks on this waterway to five and the completion of the Anderton Lift.

Later improvements include the installation of swing bridges powered by electricity, which was an innovation introduced by John Arthur Saner. The first was Hayhurst Bridge at Northwich, which opened for road traffic in 1898.

Edward Leader Williams became involved with the planning and engineering of the Manchester Ship Canal that effectively replaced the Mersey & Irwell Navigation with a new canal, which was deep enough and wide enough to handle seagoing vessels and bring them to a new dock complex at Trafford Park, Manchester. Construction of this major work was conducted between 1887 and 1894. It was opened in stages progressing from Eastham, where it joined the Mersey, through another three locks at Latchford, Irlam and Mode Wheel. Construction involved extensive excavation and some embanking. Roads were carried by swing bridges across the waterway and railways raised on embankments and tall bridges in order to provide sufficient head room for seagoing vessels to pass underneath. During construction, a temporary port was made at Saltport, at the entrance to the Weaver, to enable the trans-shipment of salt into seagoing vessels.

In sharp contrast to these major construction projects, a canal merger created a network of waterways of varying ages and standards that had to adapt to increasing competition. This was the Shropshire Union Railway & Canal Company, a new company formed to take over the Birmingham & Liverpool Junction Canal, Ellesmere & Chester Canal, Shropshire Canal and Shrewsbury Canal.

One aim was to build railways from Shrewsbury to Stafford, but they also inherited a canal network for which it was decided to operate a comprehensive fly boat network, complete with the attendant infrastructure of stables and warehouses. Many of these structures remain along the side of the navigation.

Various improvements were made at Ellesmere Port, which included the provision of hydraulic systems for the cranes and lift bridges and also the laying of railway sidings that connected with the London & North Western/Great Western Railways joint line from Hooton. With the construction of the Manchester Ship Canal, the entrance to Ellesmere Port was altered to join the ship canal, but a lock was provided for craft to cross over and pass onto the tidal River Mersey.

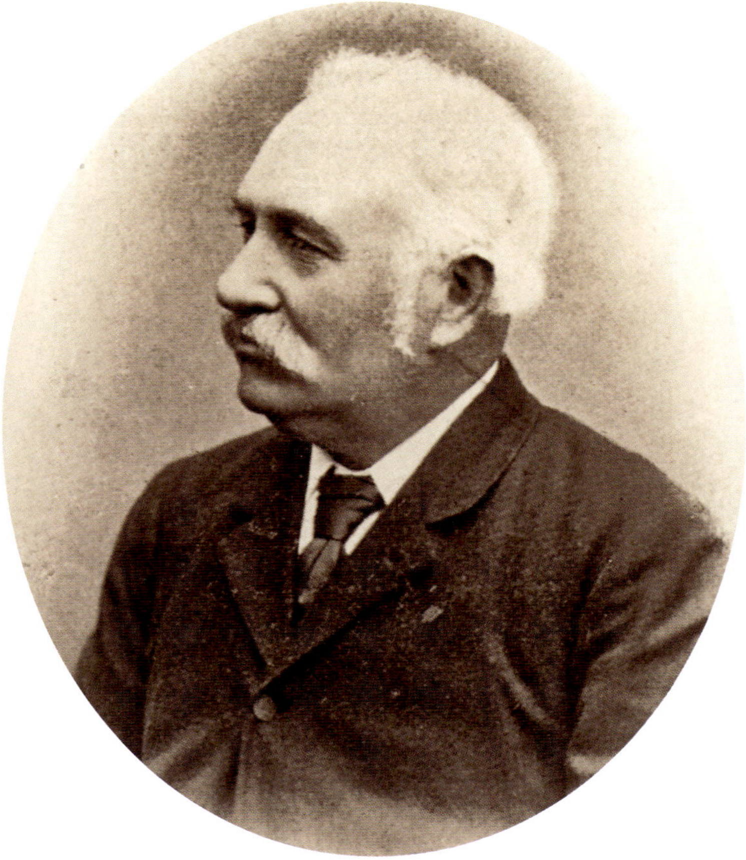

Edward Leader Williams

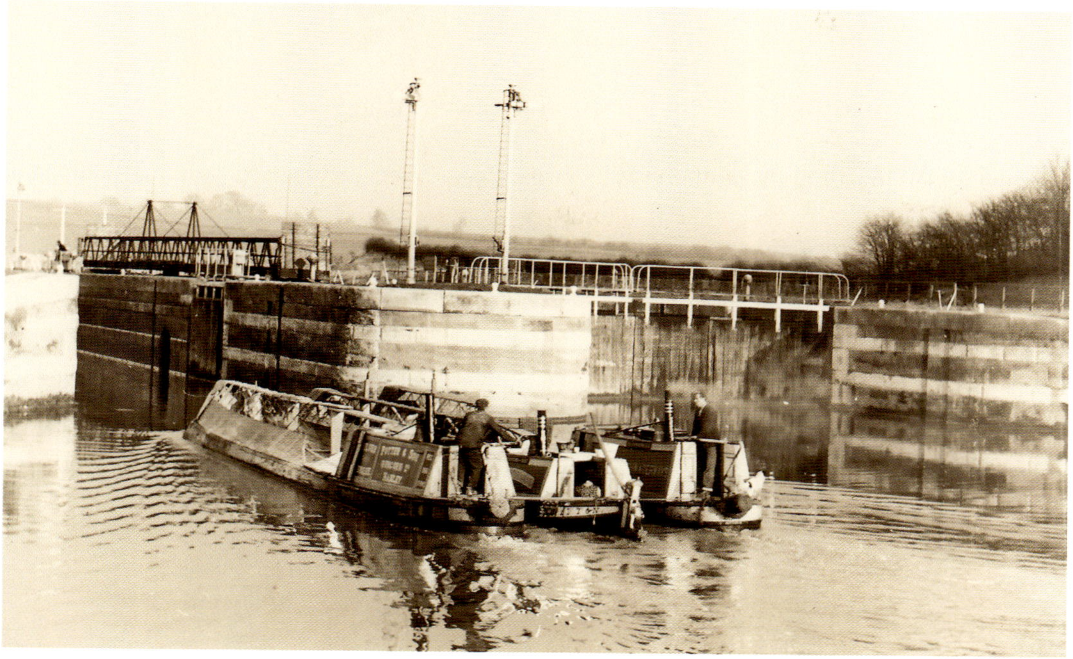

Dutton Locks, River Weaver, 1957 & 2010

Three narrow boats, two motor boats and a butty boat are seen entering Dutton Lock. Dutton became one of four locks on the Weaver under the last modernisation period from 1865 to 1890, when longer lock cuts and weirs raised the levels to suit the arrangement of a deeper, wider navigation. Locks were made from local Runcorn Sandstone and a rubbing course of Anglesey Limestone. (*Above: RCHS Hugh Compton Collection, P. Norton. Below: Ray Shill*)

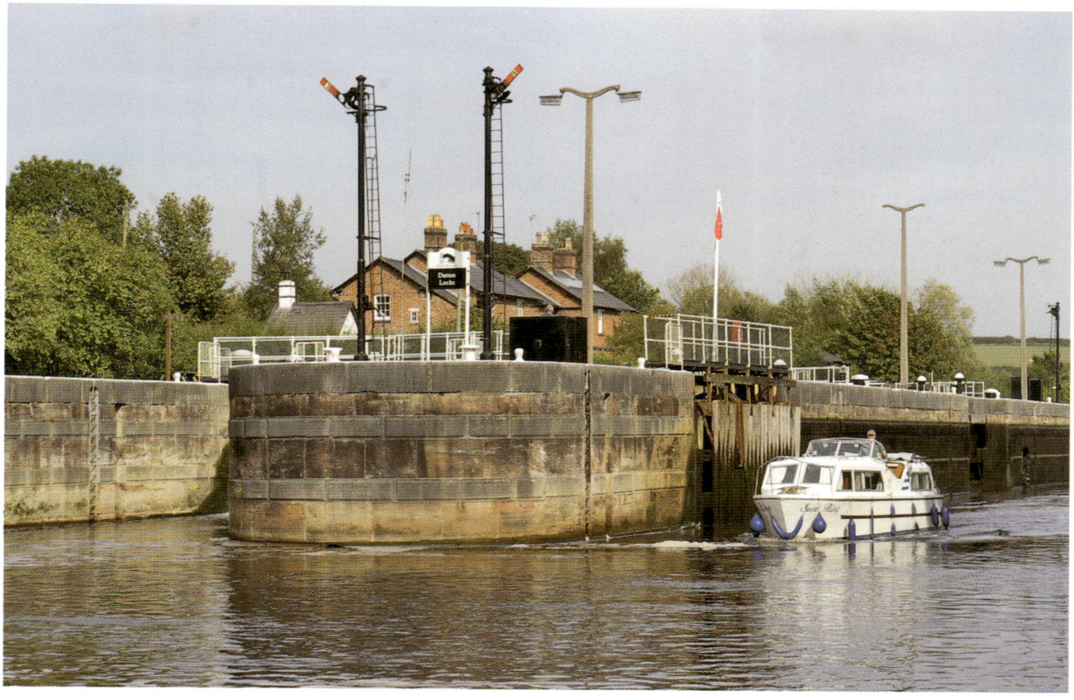

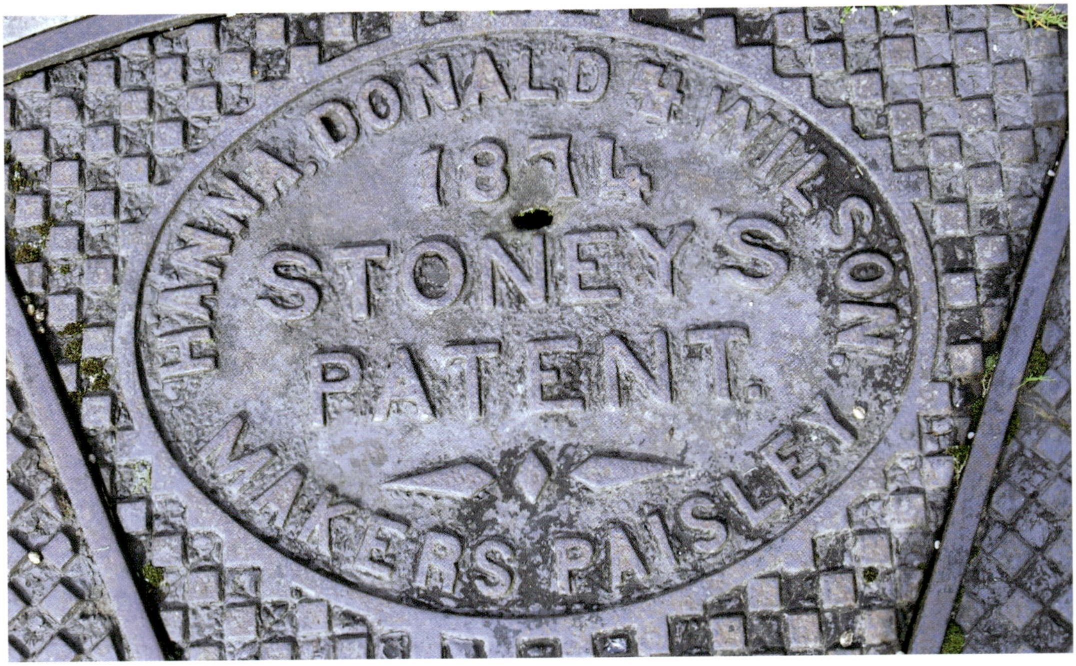

Dutton Locks, Patent Lock Mechanism, River Weaver

A feature of the new Weaver Locks was the use of pioneering technology. They were fitted with ground paddle gear designed by Francis Goold Moroney Stoney. The design incorporated turbine-driven winch gears to open and close the gates. The Oldham Ironworks Company Ltd won the contract to erect turbines and wheels at Saltersford and Dutton Locks in 1875, although this work appears to have been subcontracted to the Northmoor Foundry Co. (*Ray Shill*)

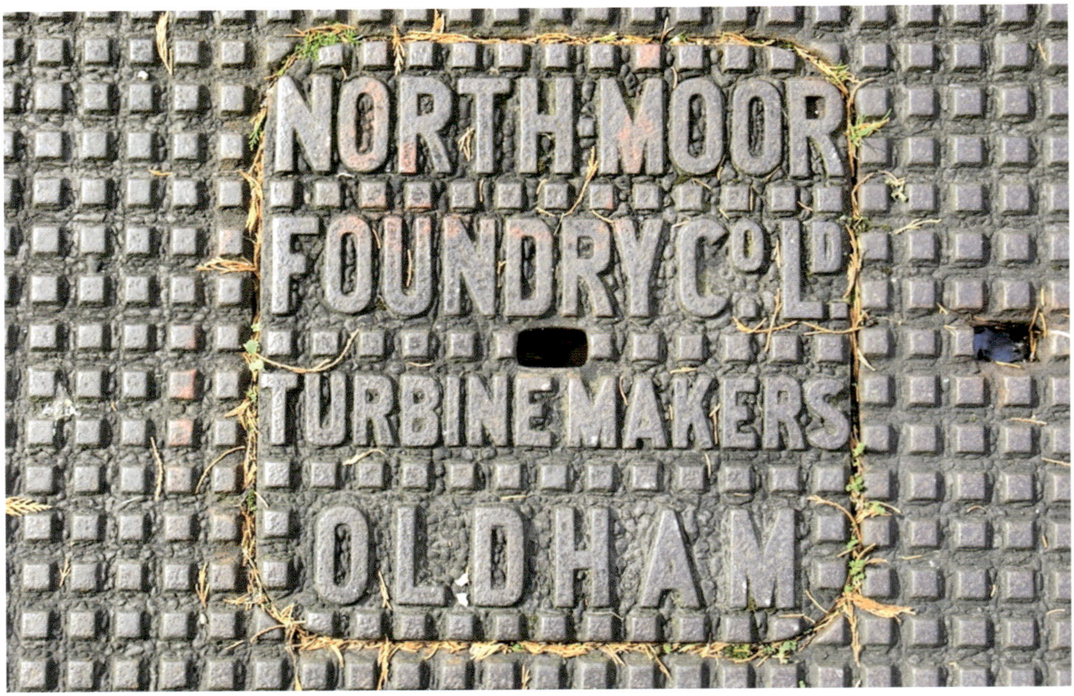

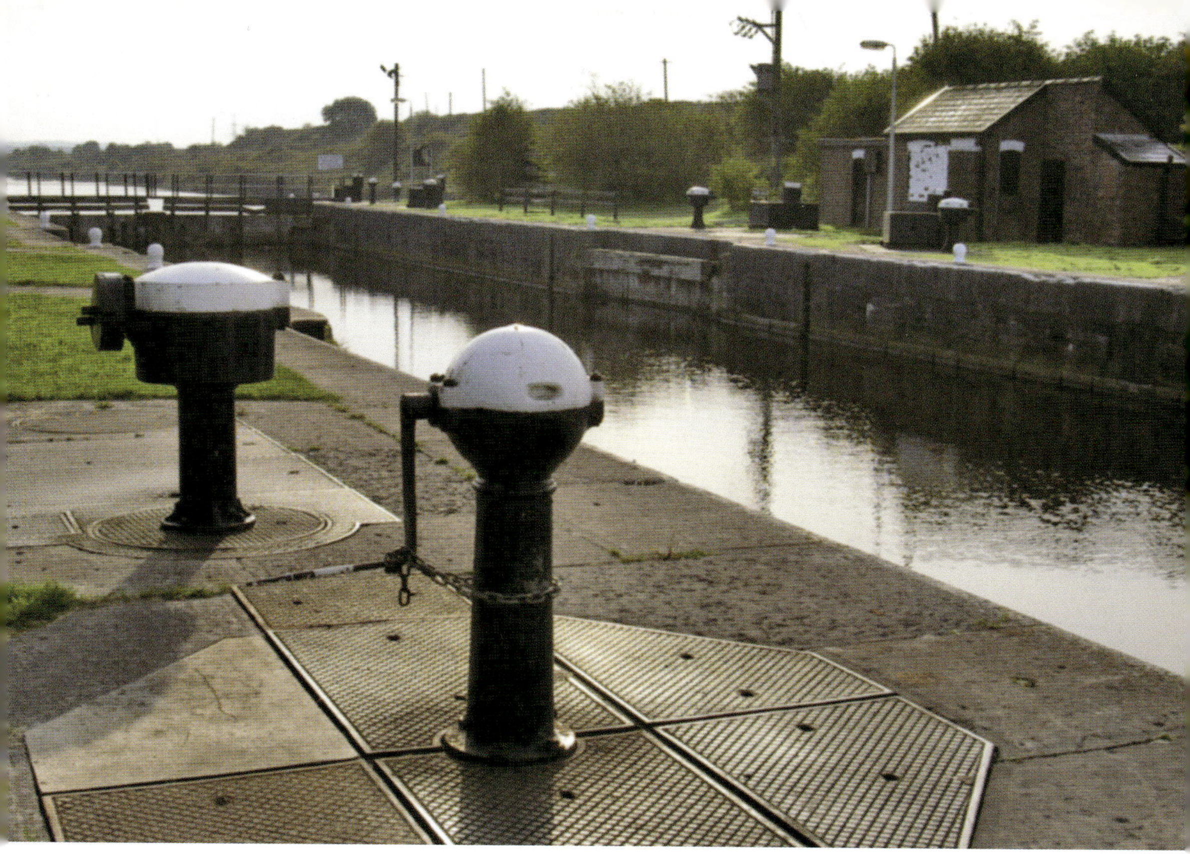

Weston Marsh Lock, Weaver Navigation
Weston Marsh Lock enables craft to pass between the Manchester Ship Canal and the Weaver Navigation. (*Ray Shill*)

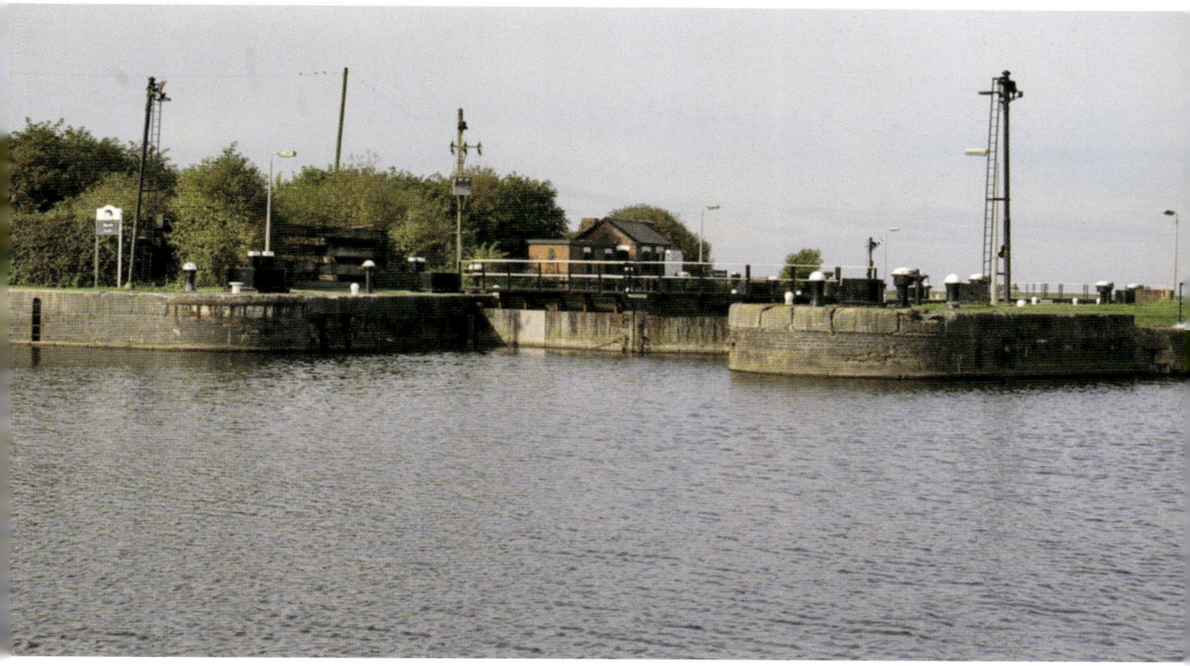

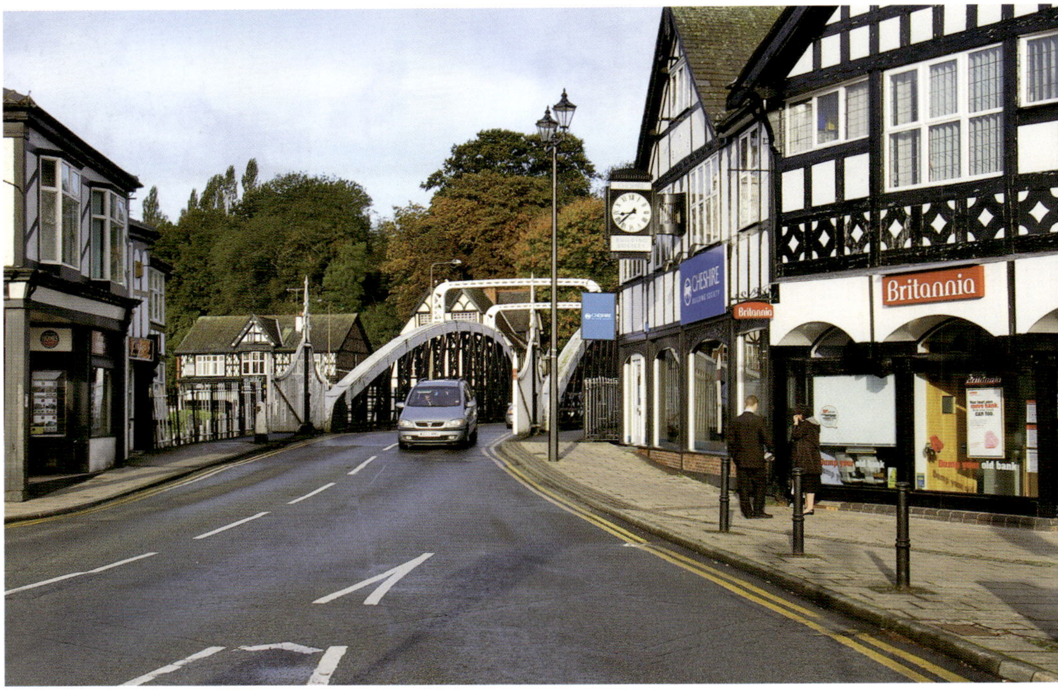

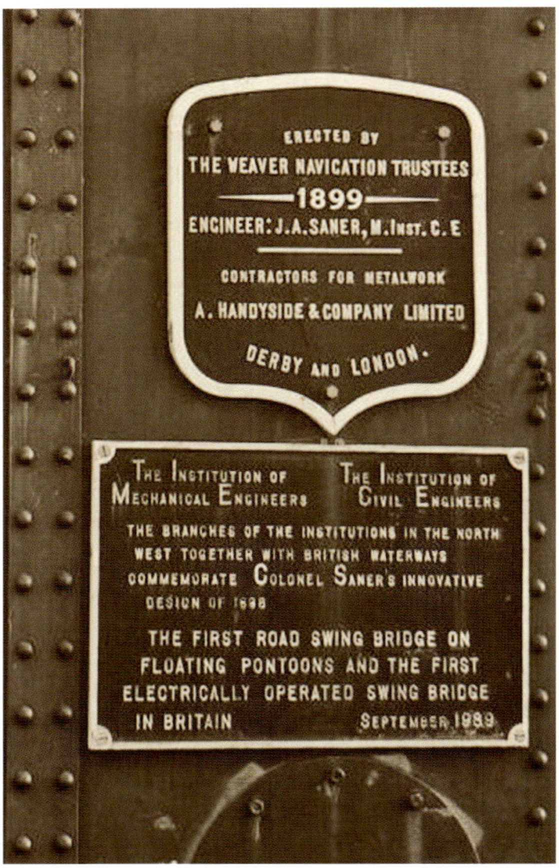

Swing Bridge Northwich and Detail of Plaque, River Weaver

Northwich Town Swing Bridge was an innovative design of 1899 which has been commemorated by plaques. These bear the name of company engineer, Saner, and maker Handyside of Derby. (*Ray Shill*)

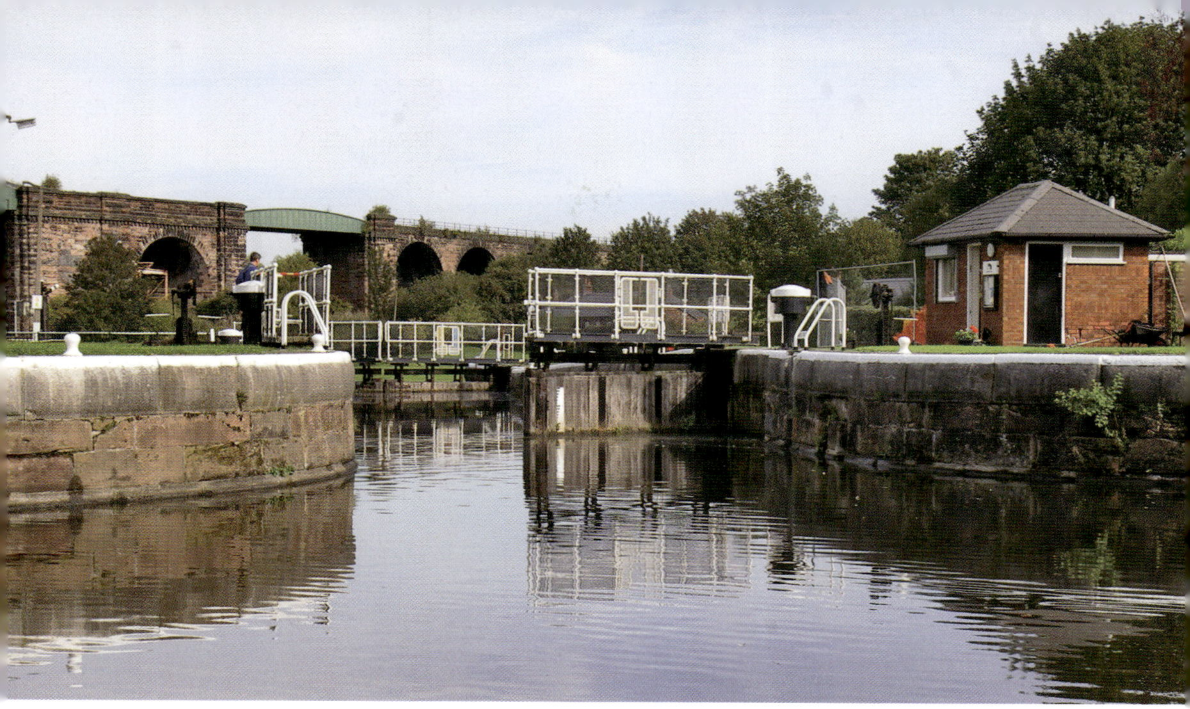

Hunt's Locks, Northwich, River Weaver

Northwich Locks are placed close to the Cheshire Lines Committee Railway that crossed the Weaver here by viaducts and bridges that were completed by 1873. The locks had various alterations; the original single lock had been rebuilt in 1847 and, following improvements instigated by William Cubitt for paired locks (1843–63), a parallel lock was made in the early 1860s. Under the revised lock scheme instituted by Edward Leader Williams to accommodate seagoing vessels, in 1887 a larger lock replaced the 'west' lock of 1847. (*Ray Shill*)

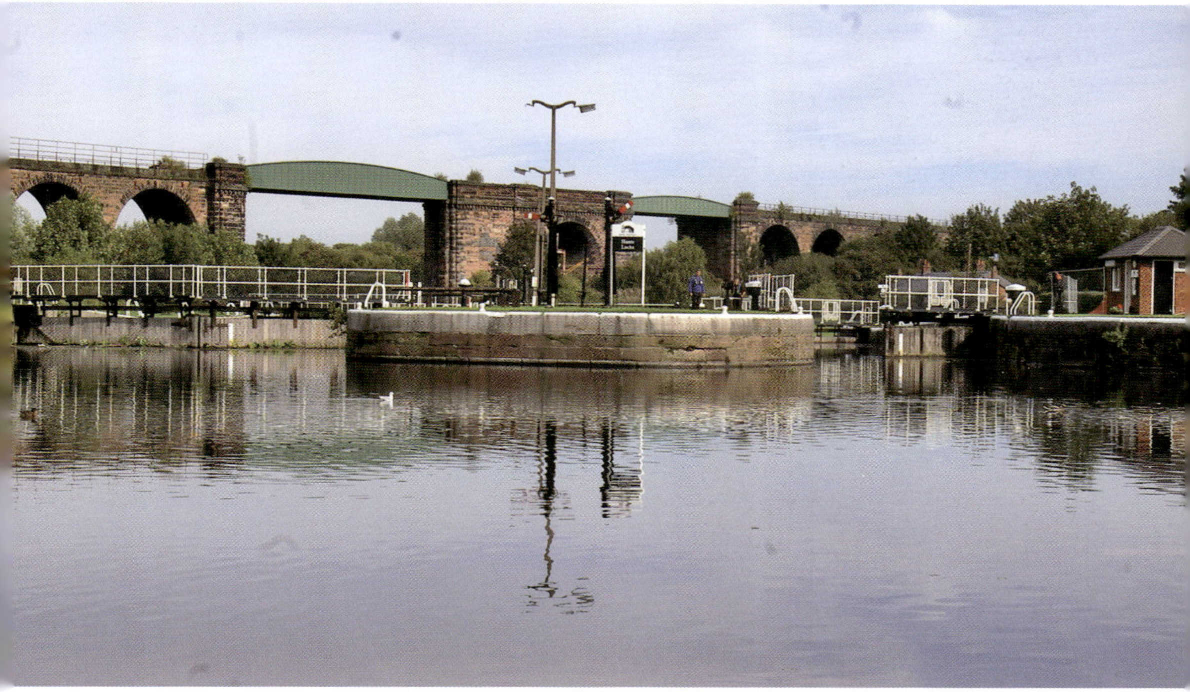

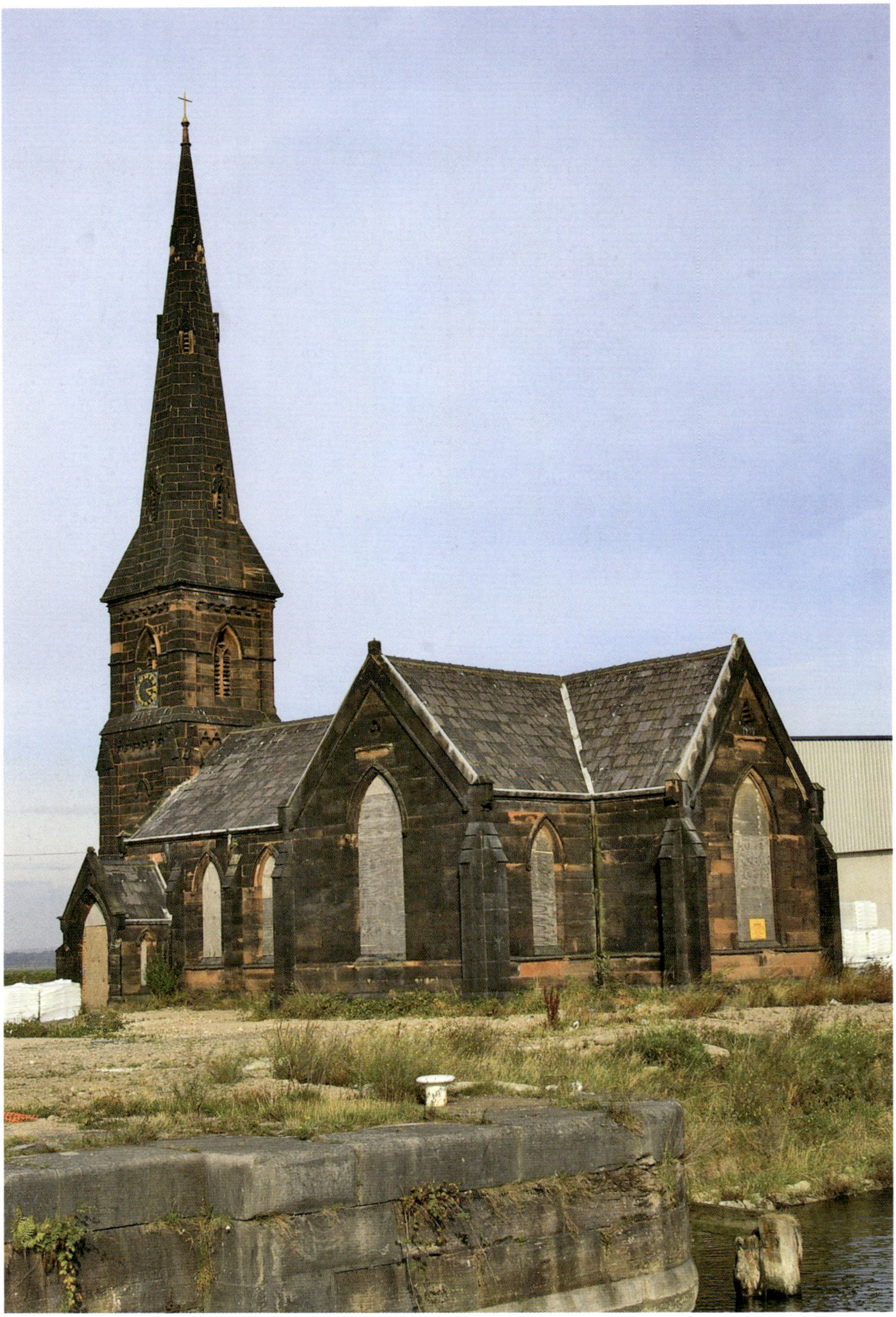

Waterman's Church, Weston Point Docks
There were three chapels that served the religious needs of the river workers such as watchmen, watermen and haulers. The Weaver Trustees built chapels at Castle Northwich, Weston Point and Winsford, once an 1840 Act of Parliament had sanctioned construction. (*Ray Shill*)

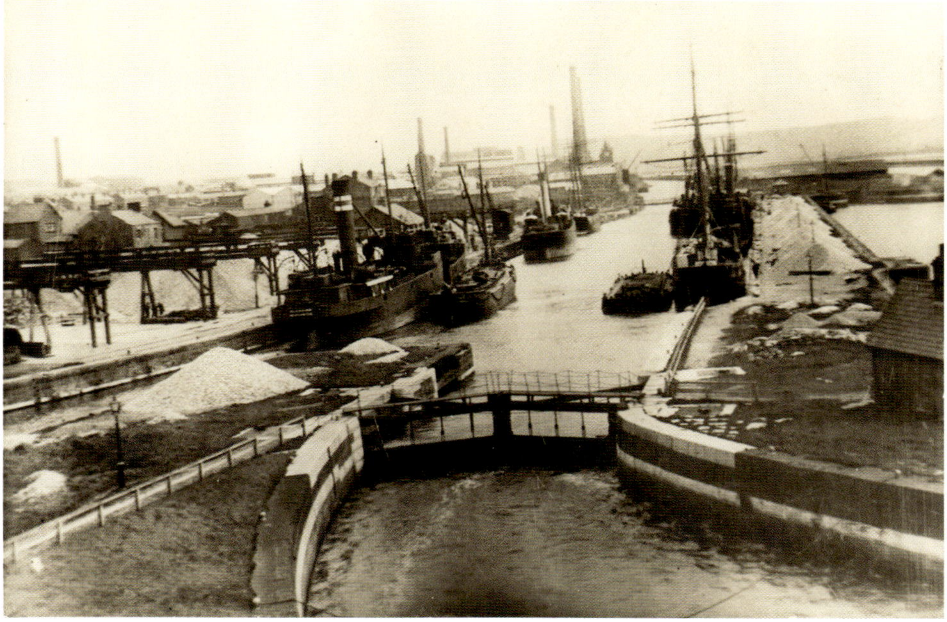

Weston Point Docks, Runcorn
The Weston Canal provided an important improvement to the Weaver Navigation that was opened from Frodsham to Weston Point in 1810. (*Railway & Canal Historical Society, Print Collection, reference 40153*)

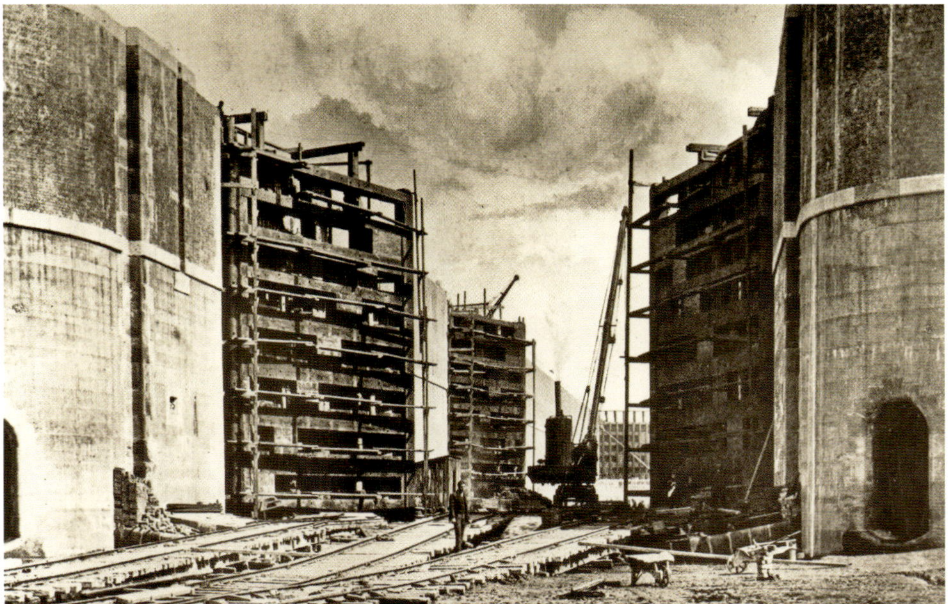

Eastham Docks, Manchester Ship Canal
Three locks were made at Eastham. The extent of their construction is evident from this image taken at the time the locks were made. (*RCHS Print Collection 40333*)

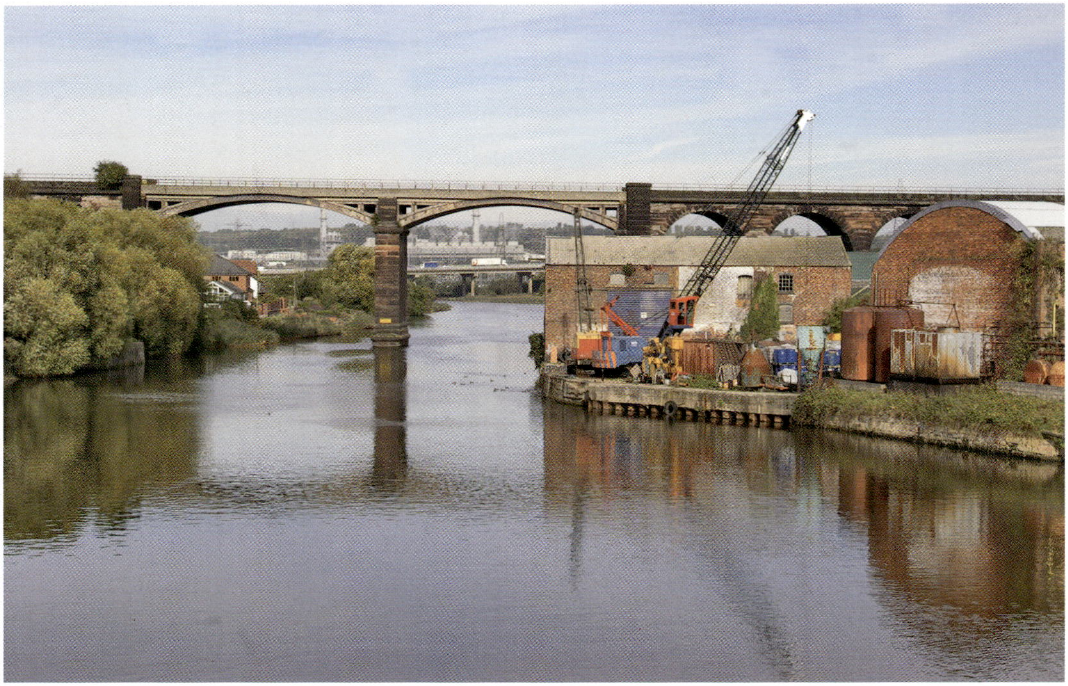

Frodsham Dock & Warehouse (*Above*) & Frodsham Dock and Buildings (*Below*), River Weaver
The original route of the Weaver Navigation was via Frodsham Lock and Bridge. From there to the Mersey, the navigation was maintained to serve mills and works on this part of the navigation. (*Ray Shill*)

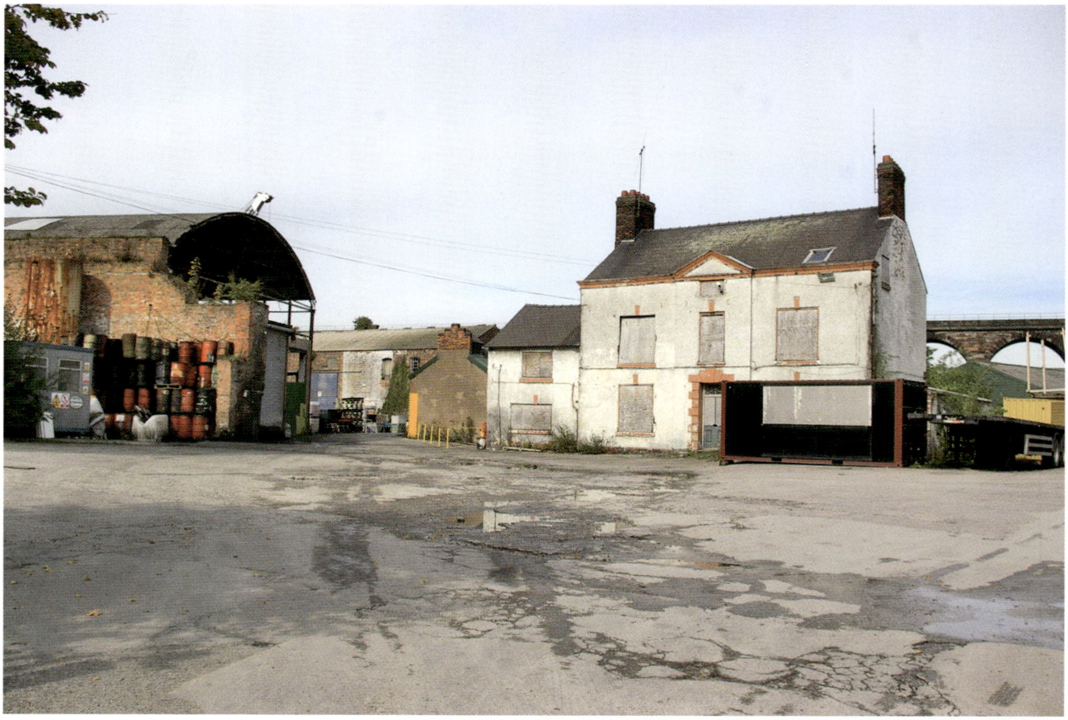

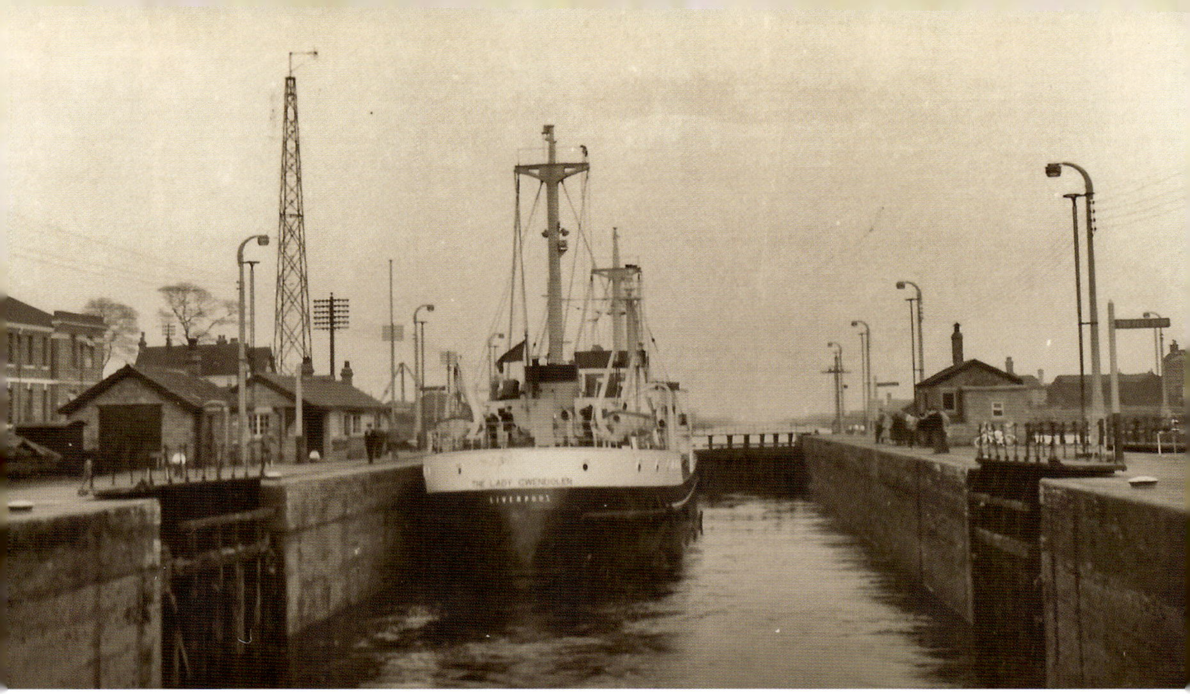

Latchford Locks, Manchester Ship Canal, 1957

The first of three locks after the Mersey Lock at Eastham was placed at Latchford near Warrington, which raised the canal to the next level that extended to Irlam. Two locks were provided here, a larger lock and a smaller lock. In this view, a Guiness ship is seen in the lock. The rise of Latchford Lock is shown in the plan reproduced from *The Engineer*, 25 May 1894. (*RCHS Hugh Compton Collection, P. Norton 65150*)

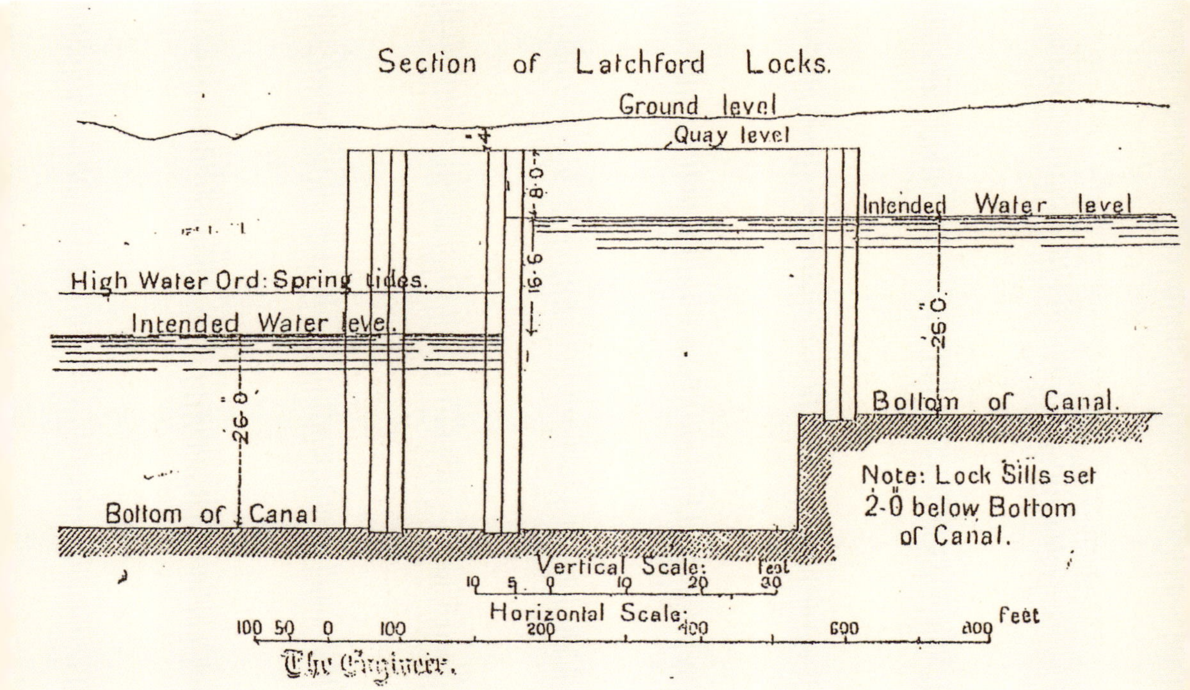

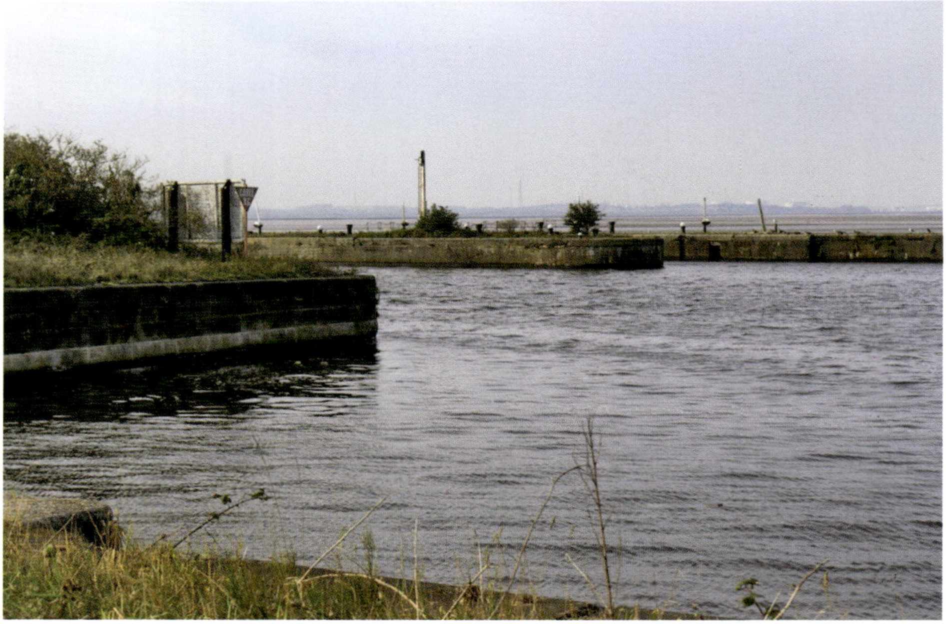

Weston Marsh Lock, Manchester Ship Canal
From Eastham to Runcorn, the Ship Canal follows the south shore of the Mersey estuary, separated from it by a stout wall. Four locks at Ellesmere Port, Weston Marsh, Bridgewater Docks (Runcorn) and Old Quay were provided to allow vessels to pass between the Ship Canal and the Mersey. This arrangement permitted craft to cross to destinations on the north bank as well as the Sankey Canal. (*Ray Shill*)

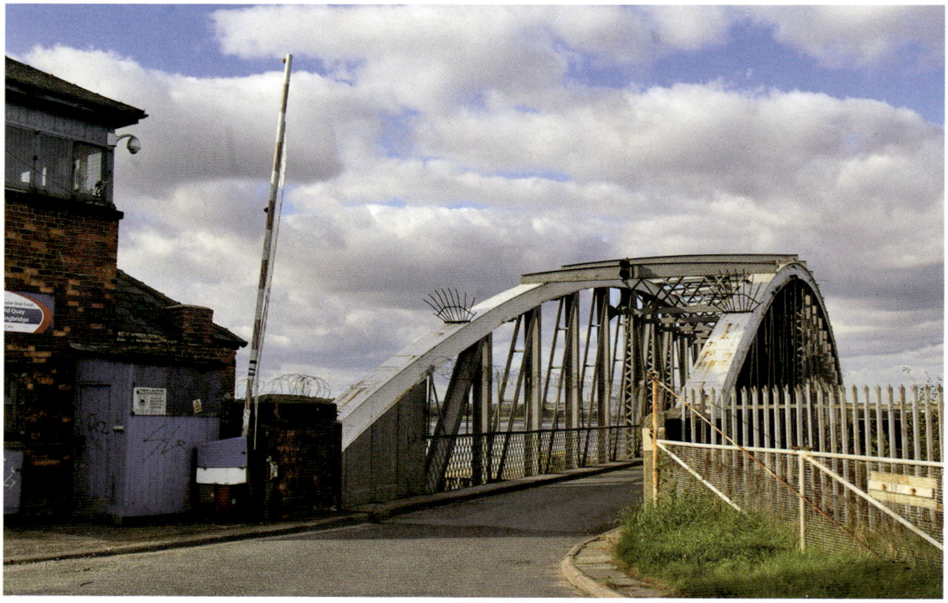

Old Quay Swing Bridge, Runcorn, Manchester Ship Canal
Old Quay Swing Bridge provides road access to a strip of land placed between the Manchester Ship Canal and the Mersey. (*Ray Shill*)

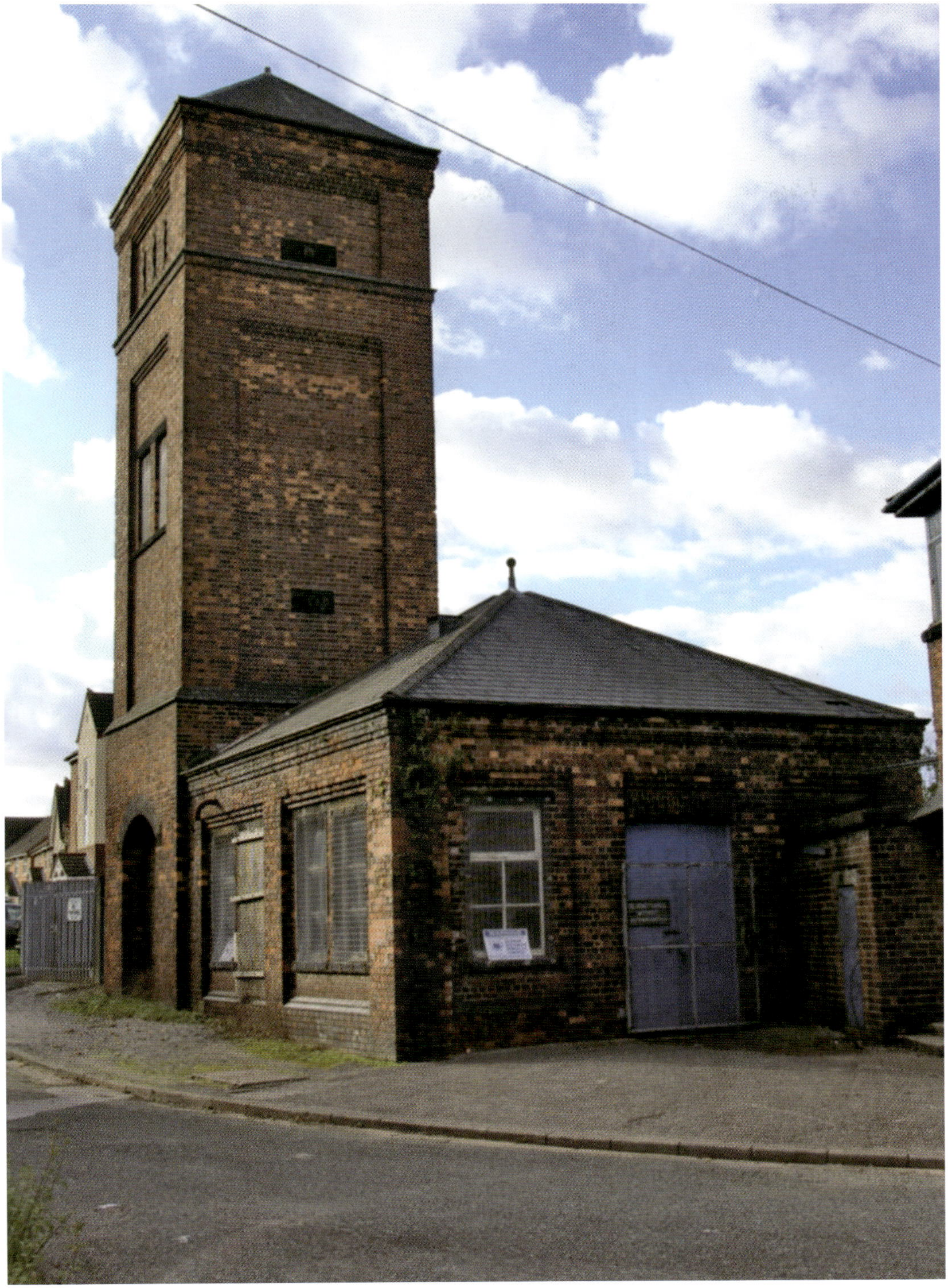

Old Quay Accumulator Tower, Runcorn, Manchester Ship Canal
A feature of the Manchester Ship Canal is the accumulator tower that assisted the opening and closing of various bridges. (*Ray Shill*)

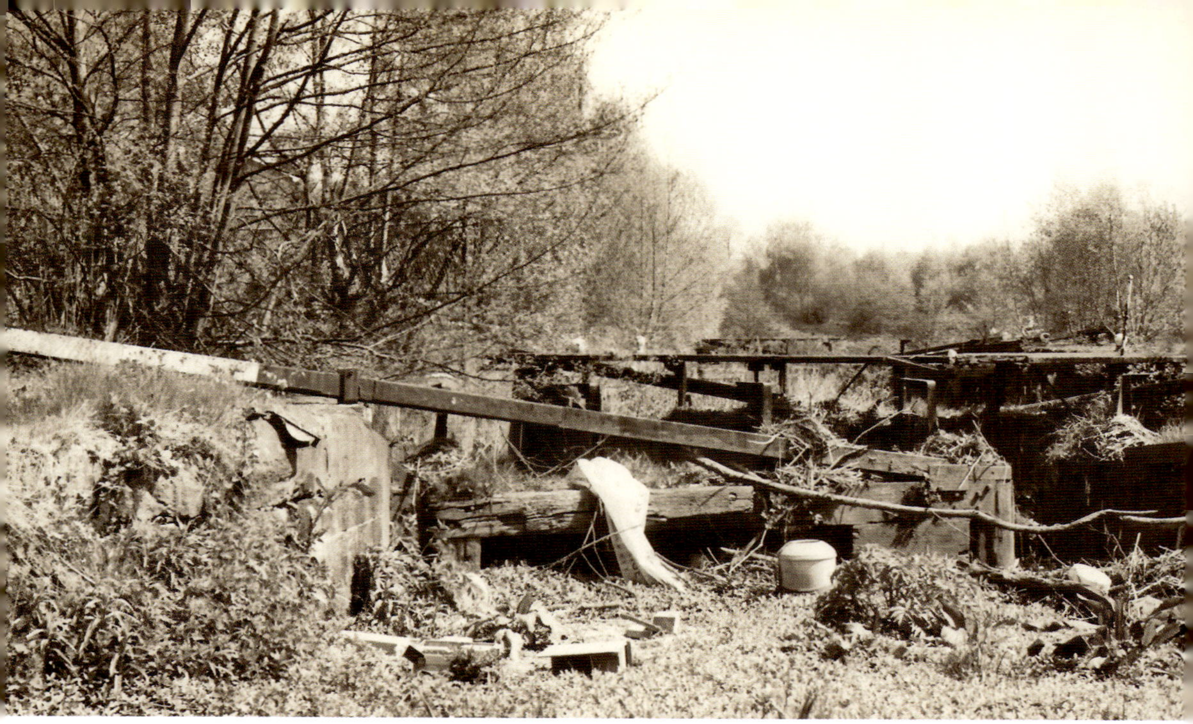

Walton Lock Mersey–Manchester Ship Canal Link, Warrington

The Mersey curved south and then north again between Arpley and Howley. When the Manchester Ship Canal was built, the west half of the curve was adapted as a link between the Mersey and the Ship Canal through the creation of Walton Lock. The remainder of the river course was filled in, as a new, straighter cut for the Mersey was made between Arpley and Howley. (*Ray Shill*)

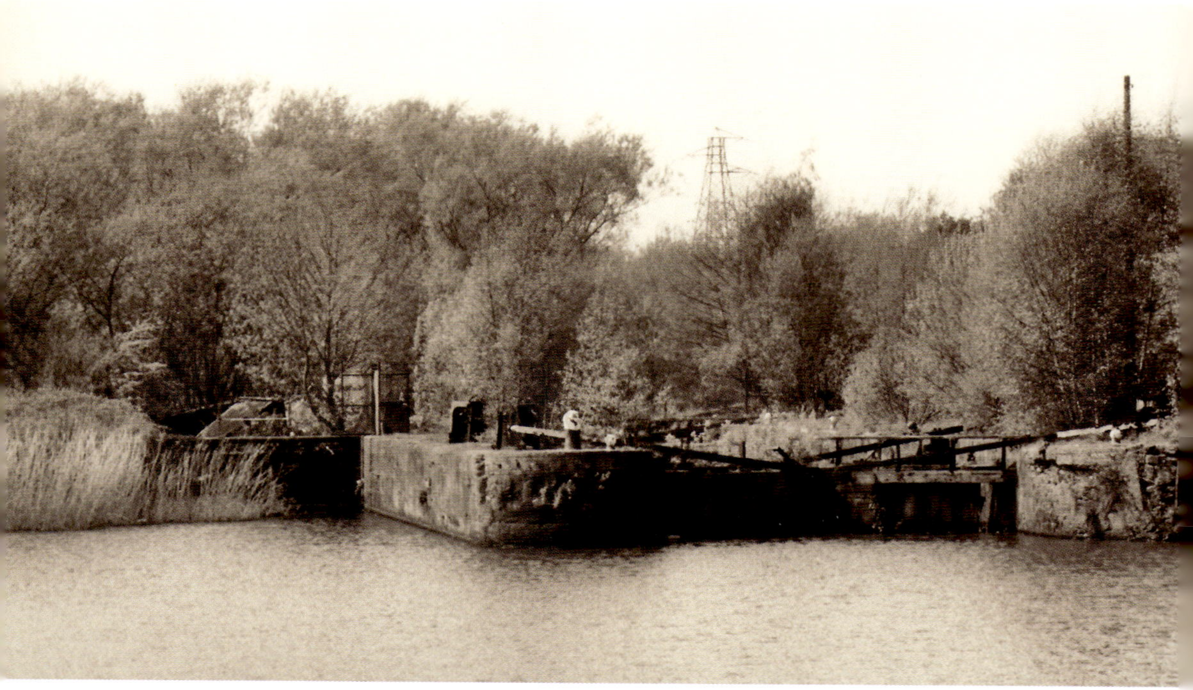

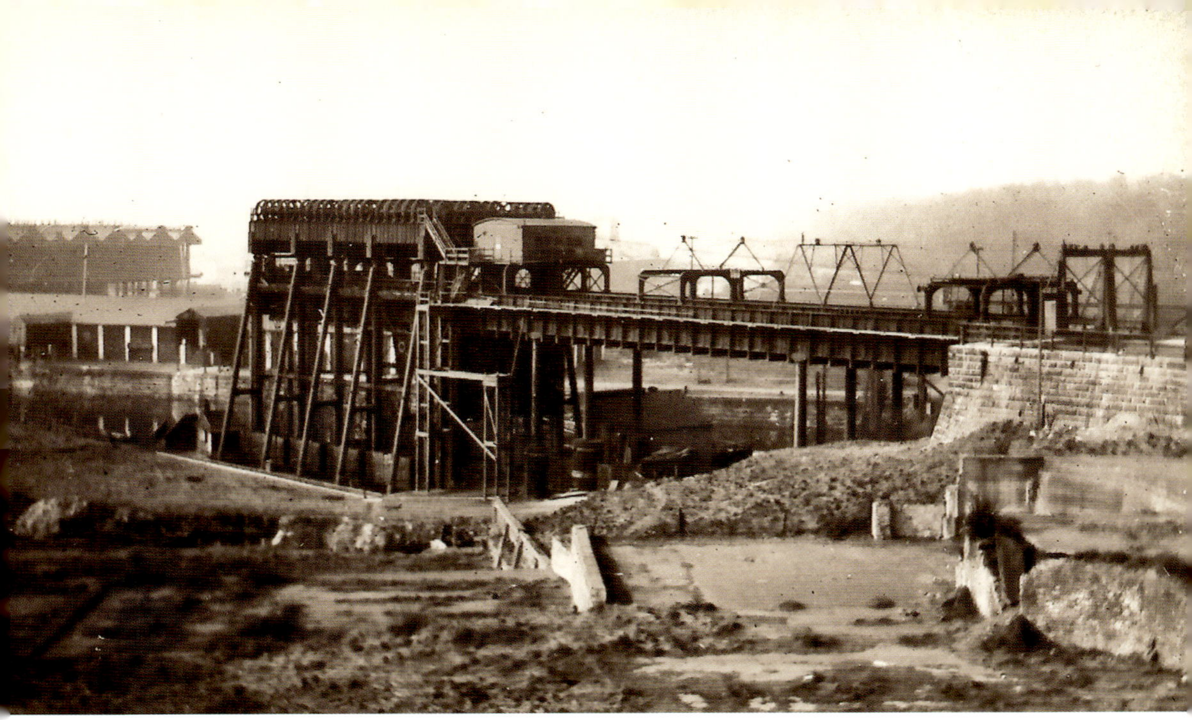

River Weaver, Anderton Lift

Improvements continued to be made to the Anderton Lift, with the hydraulic mechanism being replaced with electric haulage. Salt in the water had corroded the mechanism and the alterations were carried out to J. A. Sander's designs. (*RCHS Hugh Compton Collection 65784*)

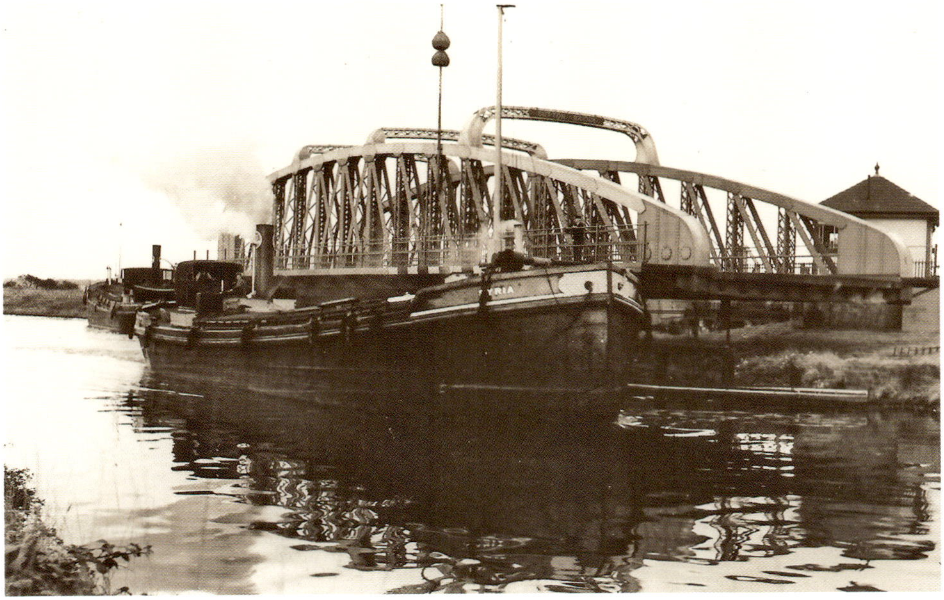

River Weaver, Frodsham

The first swing bridge powered by electricity was built at Northwich, others followed. This bridge at Sutton Weaver was constructed during 1923. (*RCHS Hugh Compton Collection 65899*)

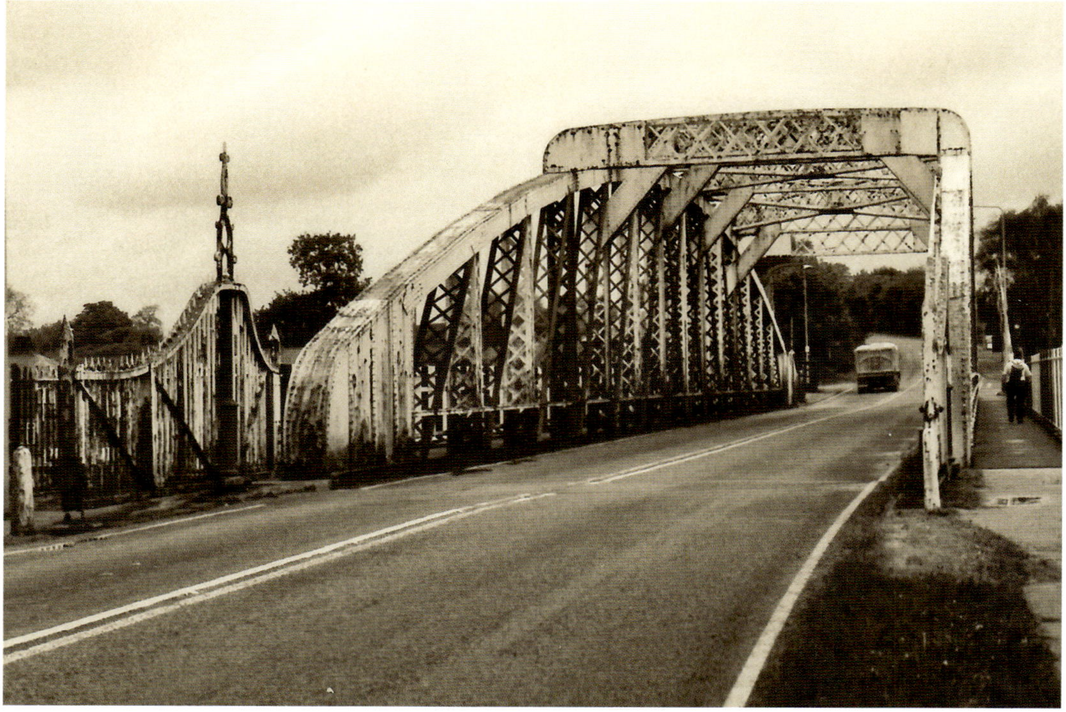

Acton Swing Bridge, River Weaver

The last major swing bridge on the Weaver was constructed during 1932 at Acton. (*Ray Shill*)

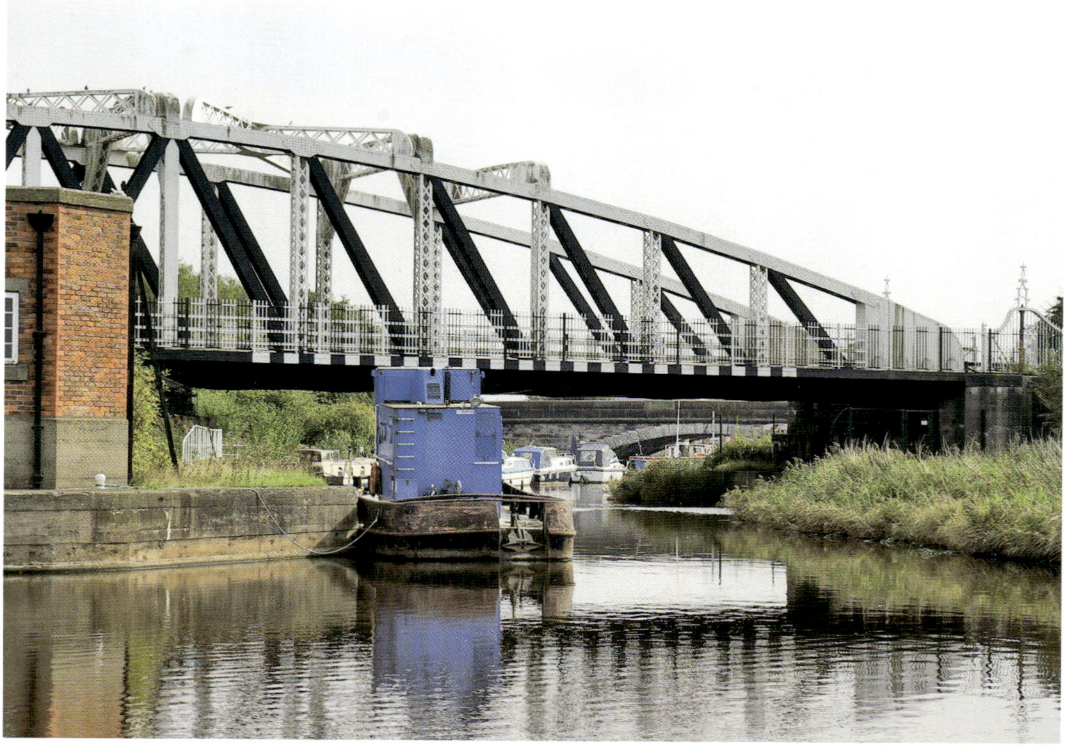

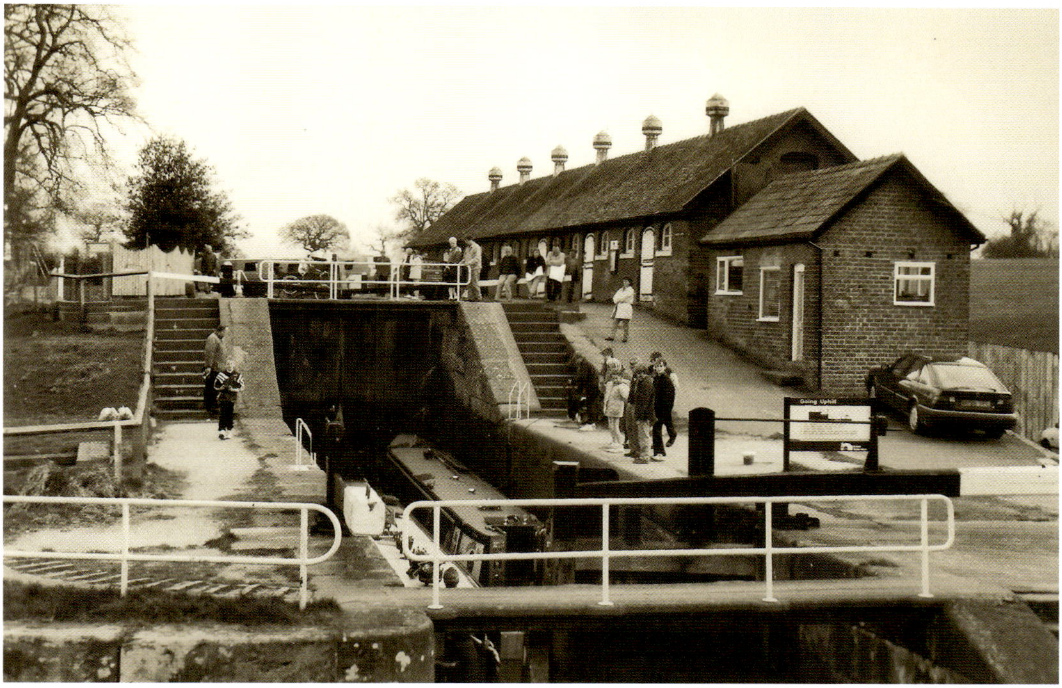

Stables and Staircase Lock, Bunbury (*Above*) & Shropshire Union Canal (*Below*)

Bunbury was an interchange point for the horses that hauled boats along the waterway for the Shropshire Union Railway and Canal Carrying Company. There was also a warehouse here for road transfer trade. (*Ray Shill*)

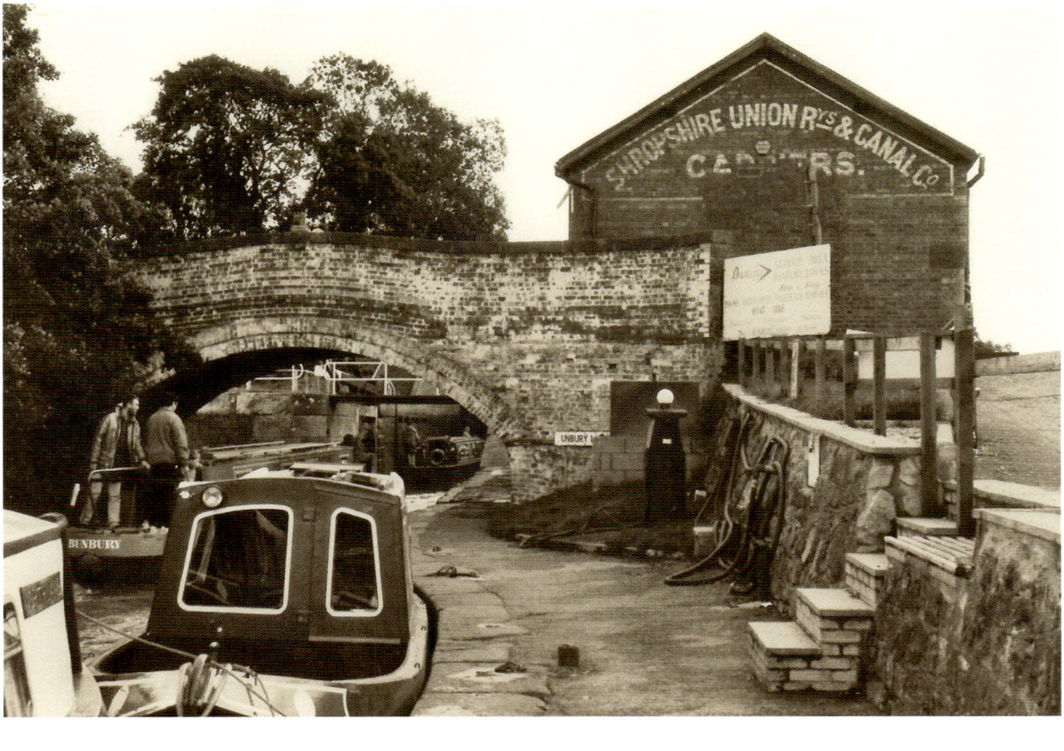

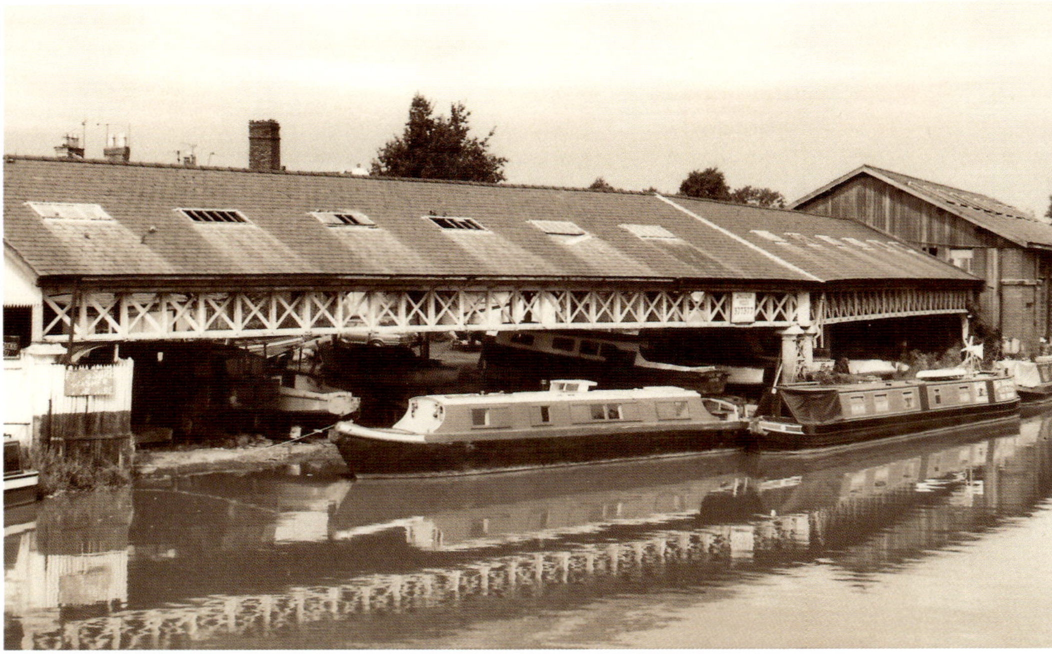

Taylor's Boatyard, Shropshire Union Canal, 1995 & 2013

This boatyard is noted for the long, overall roof that provided some protection from the elements for the boat builders and labourers who worked there. The yard was a major boatyard for the Shropshire Union Railway and Canal Company, but was acquired by Taylor in the 1930s. (*Ray Shill*)

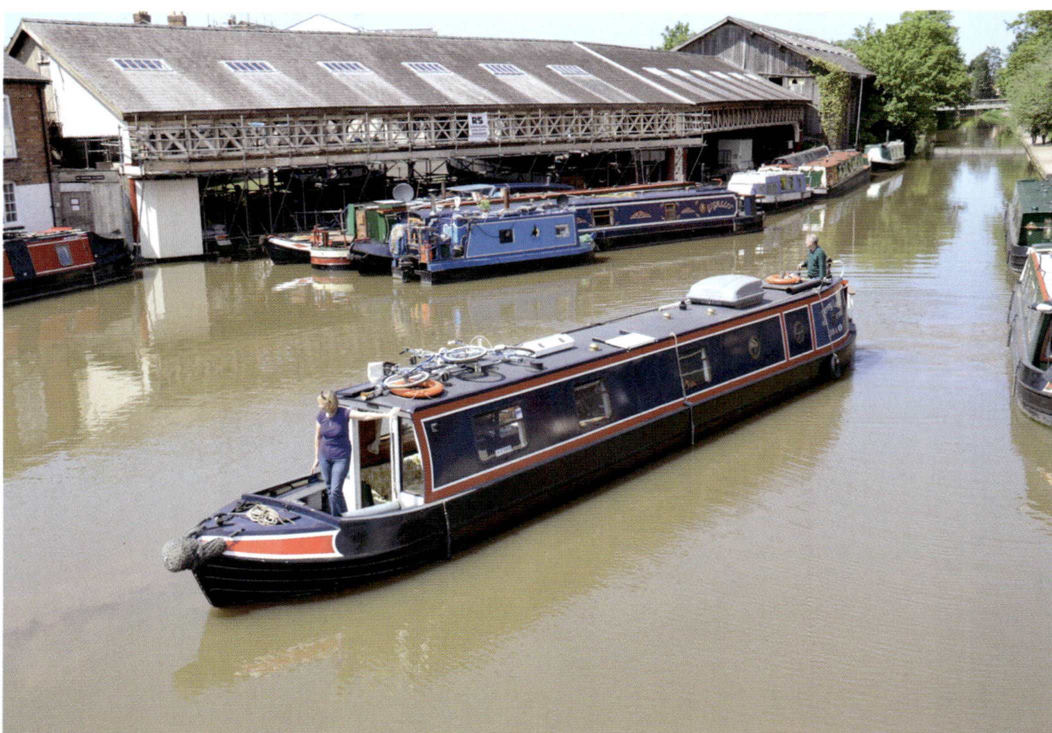

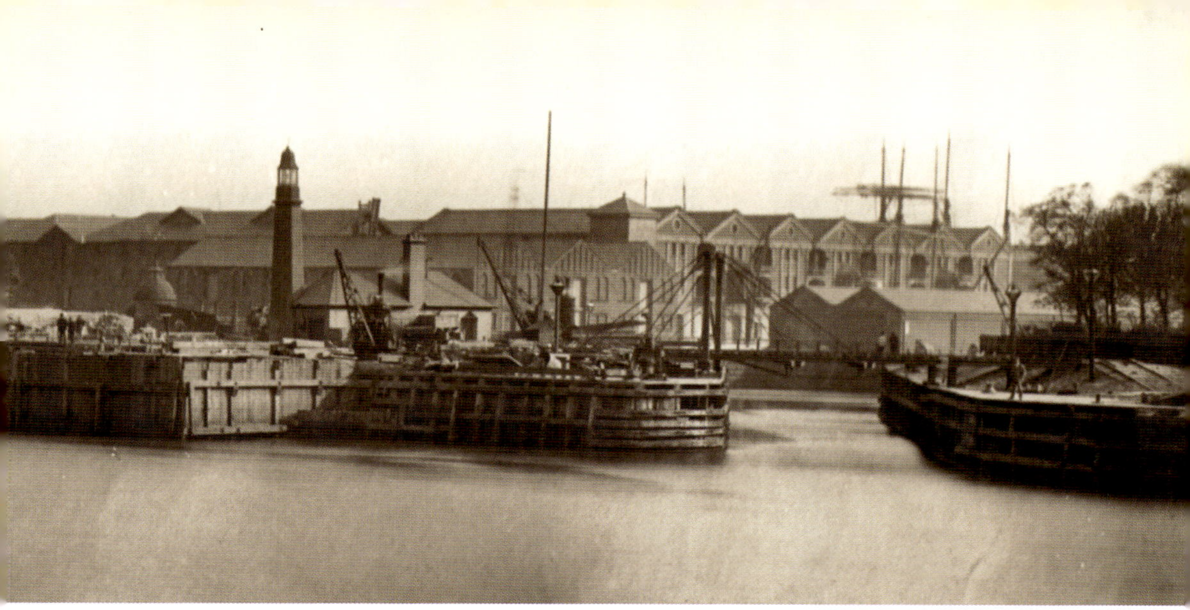

Ellesmere Port, Telford Warehouses, Lower Basin, Shropshire Union Canal, as seen from Manchester Ship Canal

The making of the Manchester Ship Canal led to reconstruction of the entrance to the Shropshire Union wharves and warehouses at Ellesmere Port. (*Railway & Canal Historical Society, Large Print Collection 82020*)

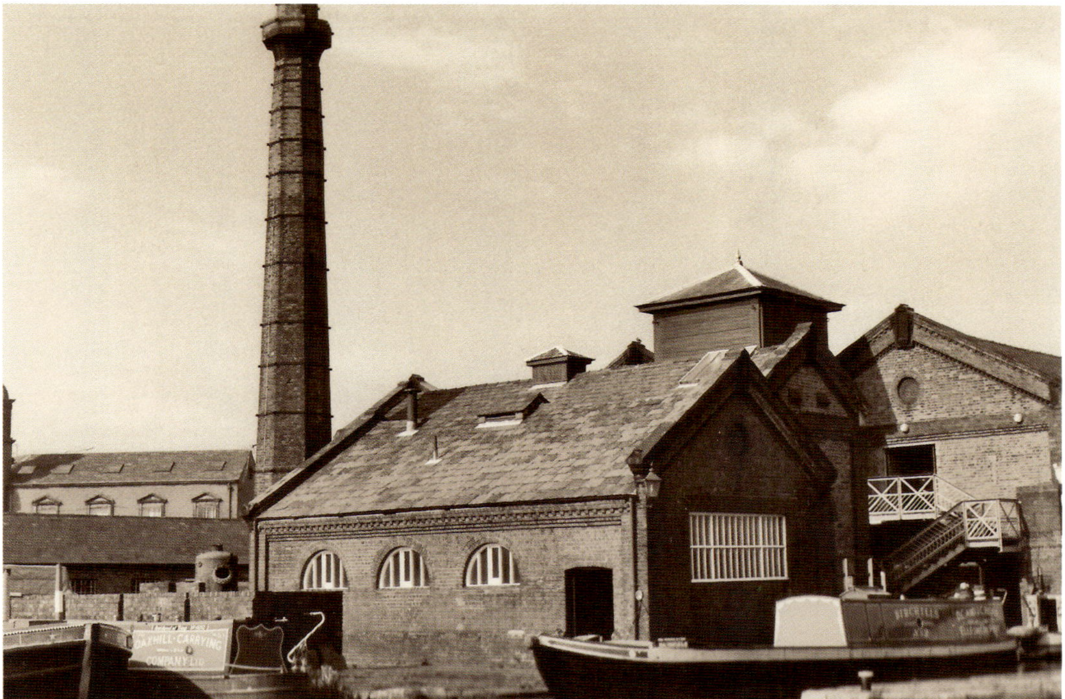

Ellesmere Port, Upper Engine House, Shropshire Union Canal

This engine house had a steam engine for the hydraulic supply to the cranes serving the warehouses at the upper wharves. (*Ray Shill*)

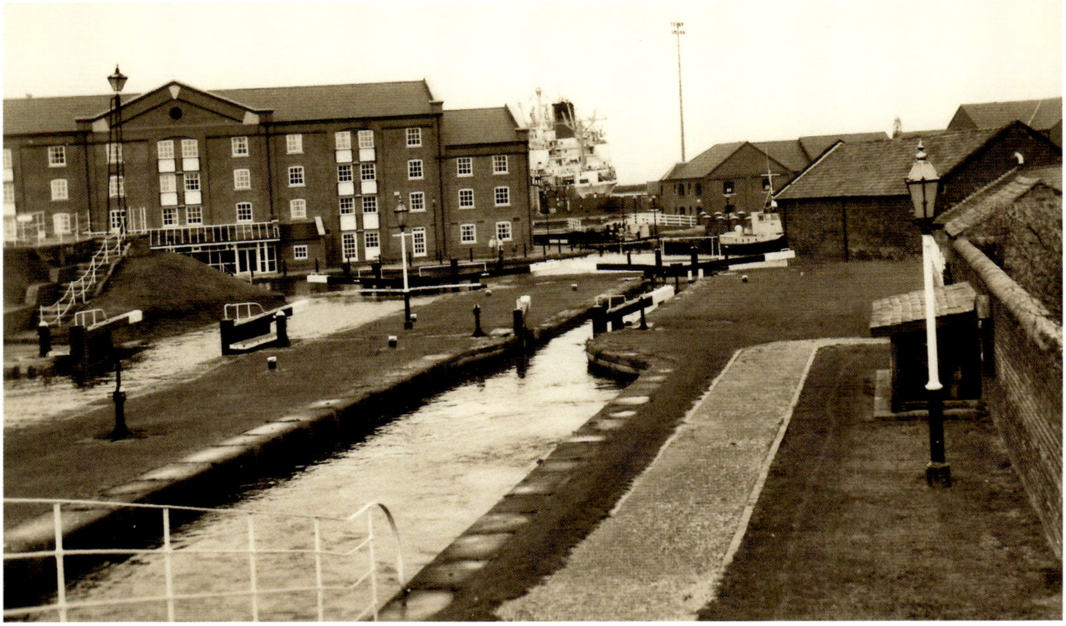

Ellesmere Port, Whitby Locks, Shropshire Union Canal
The doubling of the locks at Ellesmere Port was done to accommodate the new traffic that came on to the canal with the opening of the Birmingham & Liverpool Junction Canal between Nantwich and Wolverhampton. (*Ray Shill*)

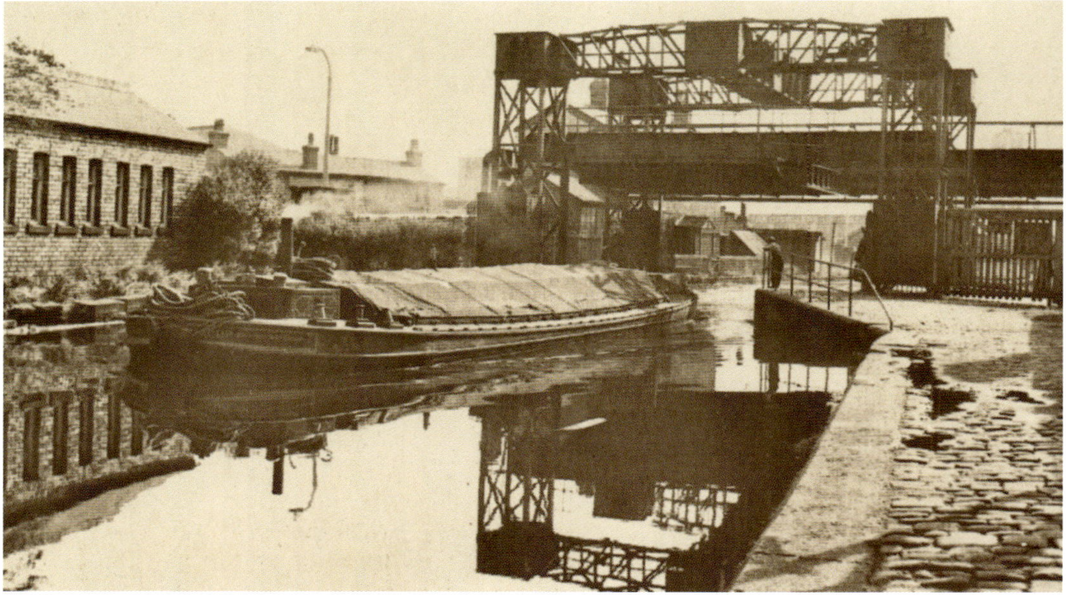

Litherland Lift Bridge, Liverpool, Leeds & Liverpool Canal
A twentieth-century innovation for the Leeds & Liverpool Canal was Litherland Lift Bridge, which was a massive structure. It replaced a swing bridge in 1934, but was demolished in 1974 when improvements for the road that passed over the bridge were required. The bridge was replaced by a modern concrete over bridge. (*RCHS Postcard Collection 96491*)

Chapter Seven

Nationalisation, Restoration & Trusts

Liverpool Docks Link, Millennium Link (Preston), Sankey Canal Restoration

Nationalisation of the waterways network in 1948 brought the Leeds & Liverpool, Shropshire Union and River Weaver under the control of the North West Division of the Docks & Inland Waterways Executive, which inherited a dwindling trade on these waterways. The DIWE handed over control to Britain's Waterways and then, from 1963, British Waterways, who dealt with the changeover from commercial to pleasure boat traffic, with perhaps less ruthless closures than were instigated elsewhere on their system. The only major closure was the final abandonment of the Sankey Canal.

A separate entity was the Manchester Ship Canal Co., which had acquired the Bridgewater Canal in 1887. They allowed the abandonment of waterways such as the Runcorn Locks and Runcorn & Weston Canal, and concentrated on trade to Runcorn Docks by the Ship Canal. Remaining segments of the Runcorn & Latchford Canal from Twenty Steps Lock to Manor Lock were closed and later filled in. New Woolston Cut and Howley Lock were similarly abandoned. They also allowed the removal of the old warehouses at Sutton Heath for new housing.

Improvements were made at the Eastham end, building a new Queen Elizabeth II dock (1949–54) and Eastham Oil terminal with a pipeline link to the oil refineries at Stanlow. Increases in the size of shipping had an effect on vessels using the canal, which declined from the 1970s, although the section to Stanlow remained busy. In recent years, new trade has returned to the canal, in particular container traffic. Peel Ports presently own the Ship Canal, Bridgewater Canal and the Port of Liverpool and from 2007 have run a container service from Liverpool to Irlam.

British Waterways were associated with two major schemes in the North West: the Millennium Ribble Link and the Liverpool Canal Link. The Ribble Link started as a volunteer effort, but was embraced by BW and was opened in July 2002. Making the link was the joint responsibility of the Ribble Link Operation and Construction Ltd and the Waterways Trust, a national charity formed in 1999. The route incorporated the Leeds & Liverpool Canal (Rufford Branch), the Douglas Navigation, Savik Brook and joined the Lancaster Canal at Ingol. Eight locks and one tidal gate were required to make the Savik Brook navigable. British Waterways' responsibility included maintaining the navigation with a particular commitment to dredging. From July 2012 this responsibility passed to the Canal & River Trust, when they took over all British Waterways navigations in England and Wales.

The Liverpool Canal Link was constructed between 2007 and 2009. It was opened officially on 25 March 2009, and commenced regular craft movements in April. The existing head of navigation from Stanley Docks and the entrance through Waterloo Dock to the Mersey was altered, making a new channel through the infilled Trafalgar Dock to Prince's Dock, and another new channel across Mann Island to Canning Dock, Albert Dock and Salthouse Dock.

While the Sankey (St Helens) Canal was closed in a piecemeal manner, the last section was closed in 1963. The lock at Fiddlers Ferry was restored by Warrington Council in the 1980s to provide access to a part of the canal to form a new marina. During 1985, the Sankey Canal Restoration Society was formed to restore the navigation throughout and has restored specific features in order to reopen navigation along the former waterway.

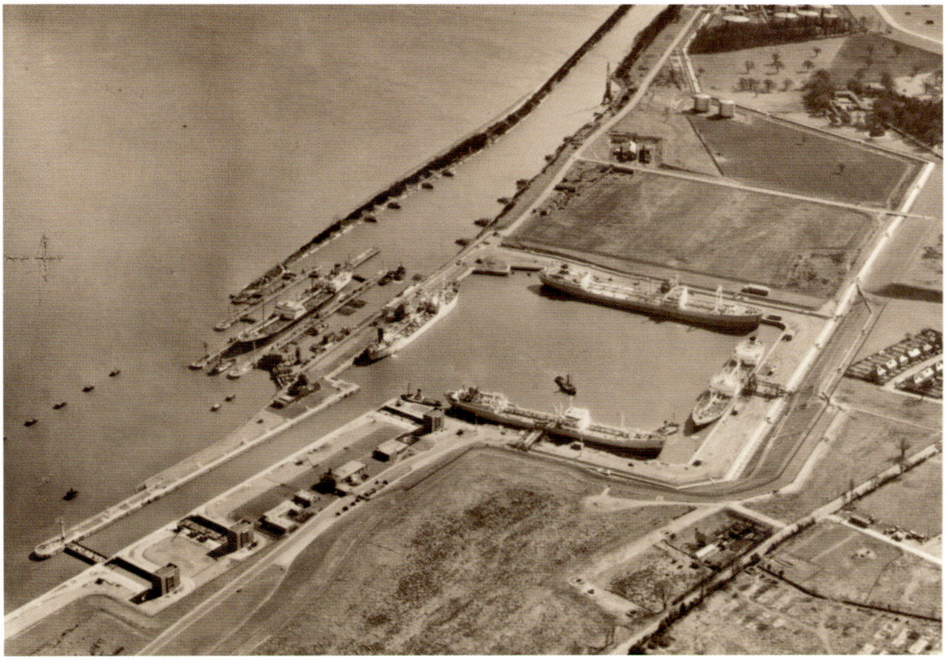

Queen Elizabeth II Docks & Eastham Locks, Manchester Ship Canal
The docks at Eastham were opened in 1954 to provide access for tankers that were too large to pass through the locks and the canal to Stanlow. (*RCHS Print Collection 40336*)

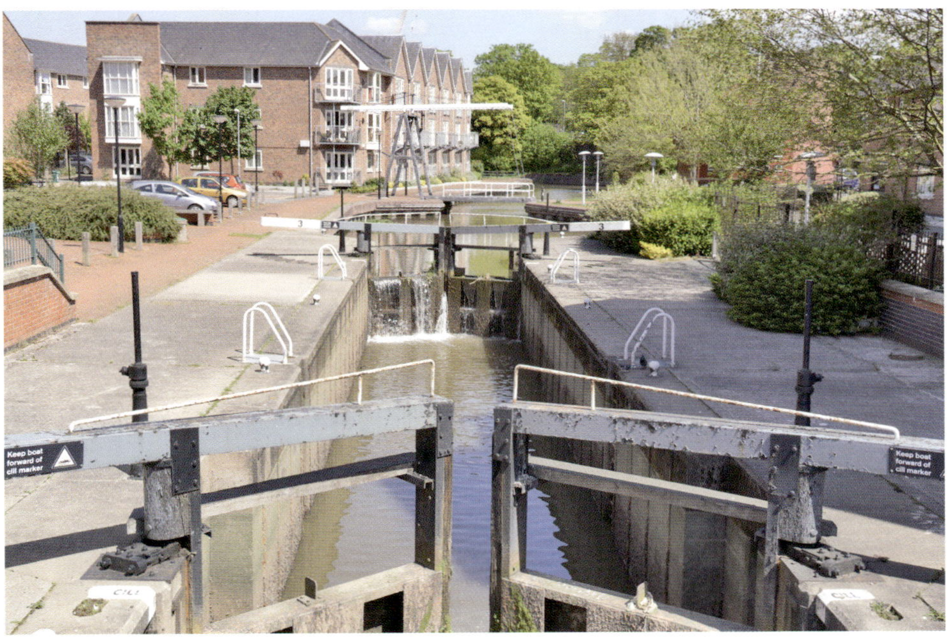

New Lock and Housing Development, Dee Branch, Chester Canal
The former basin and warehouses at Chester on the River Dee branch has been partly filled in and the land used for new housing and this new development includes a Shropshire Union style lift bridge over the waterway. (*Ray Shill*)

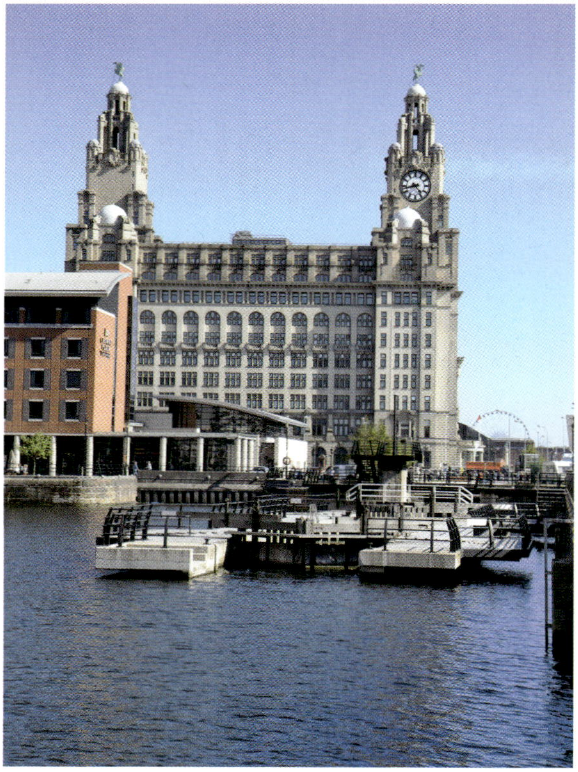

Lock Five, Liverpool Canal Link, Prince's Dock
From April 2009, a link to Salthouse Dock became navigable for craft passing down the four locks of the Stanley Lock flight from the main line of the Leeds & Liverpool Canal. Lock five enabled craft to descend to a lower canal that passed through three tunnels made in Mann Island to Lock six. (*Ray Shill*)

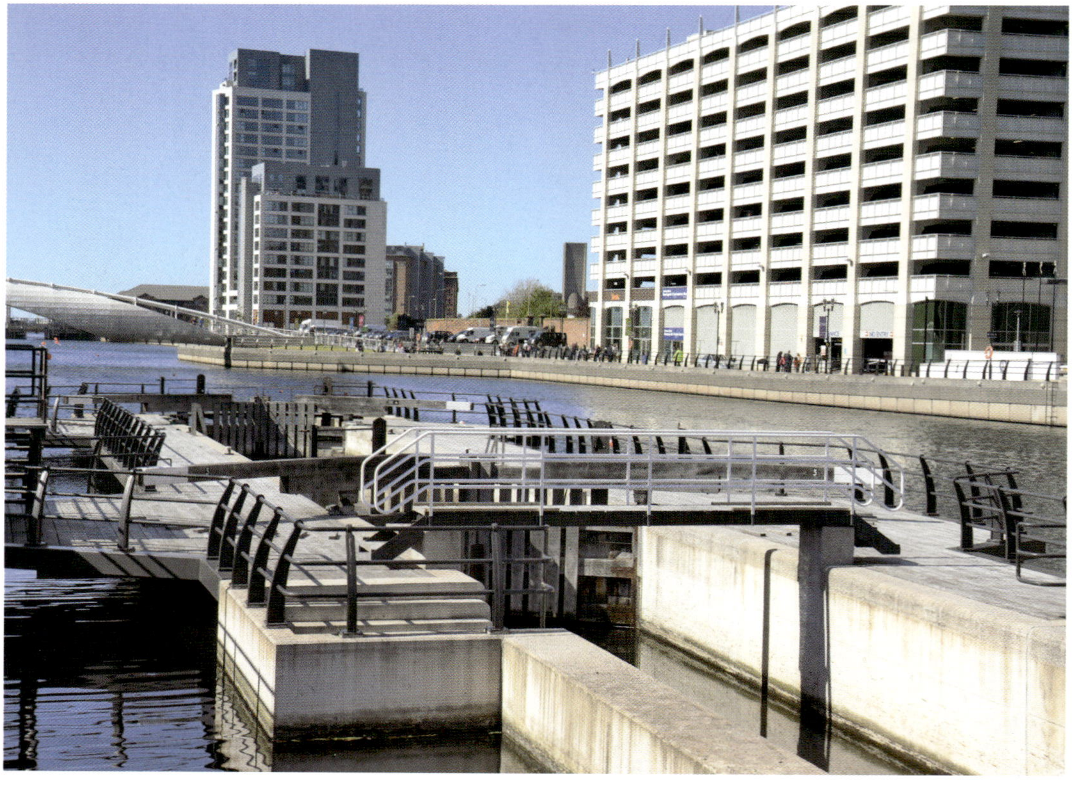

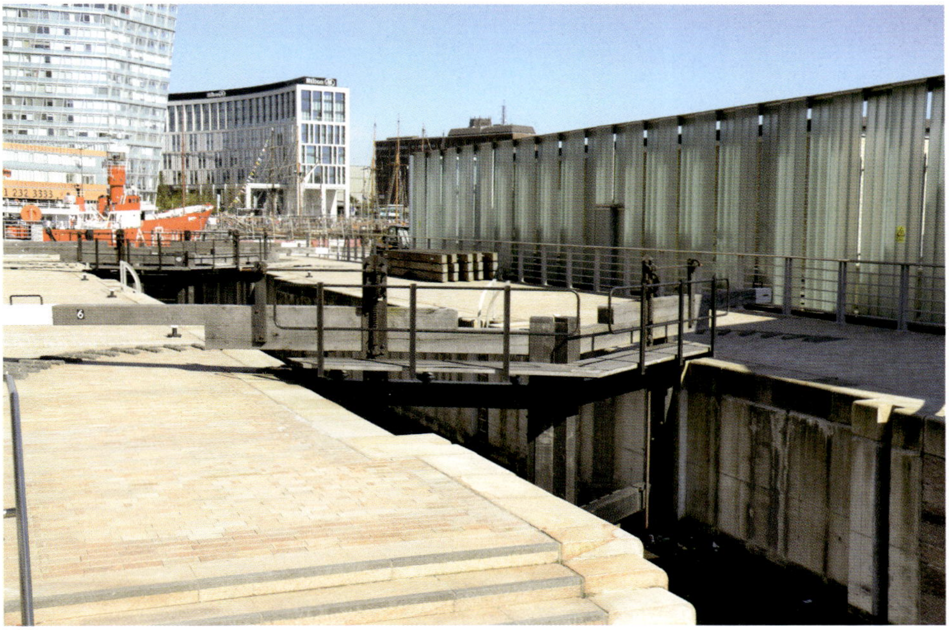

Lock Six, Liverpool Canal Link, Canning Dock

Lock six enabled craft to rise again to the normal dock level at Canning Dock. In effect, the section between Lock five and six might be considered as the lowest point of the navigation through to Salthouse Dock. (*Ray Shill*)

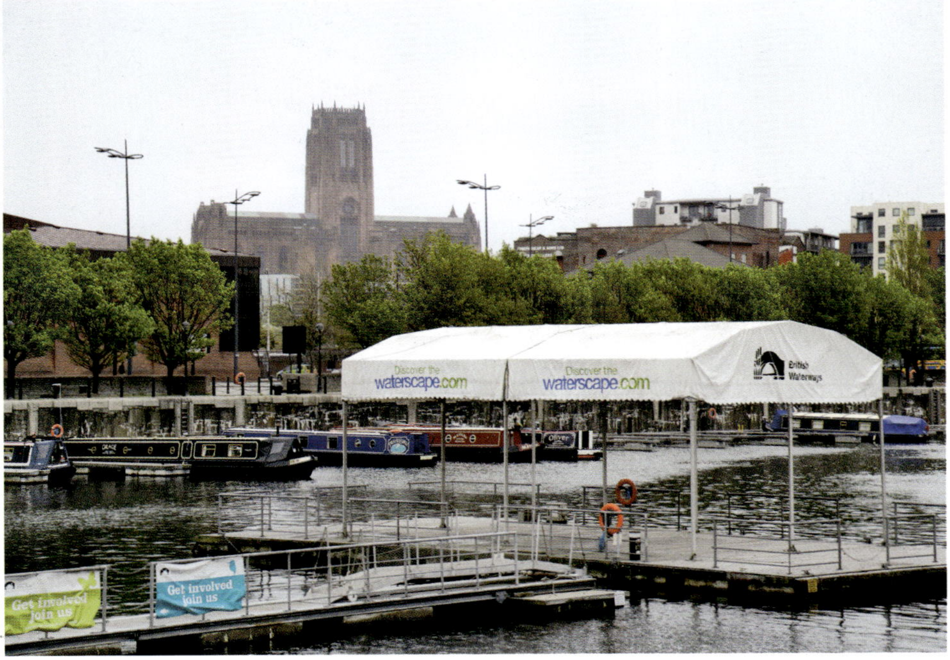

Salthouse Dock, Liverpool

Boats travelling along the Liverpool Canal Link from Stanley Docks moor at Salthouse Dock. (*Ray Shill*)

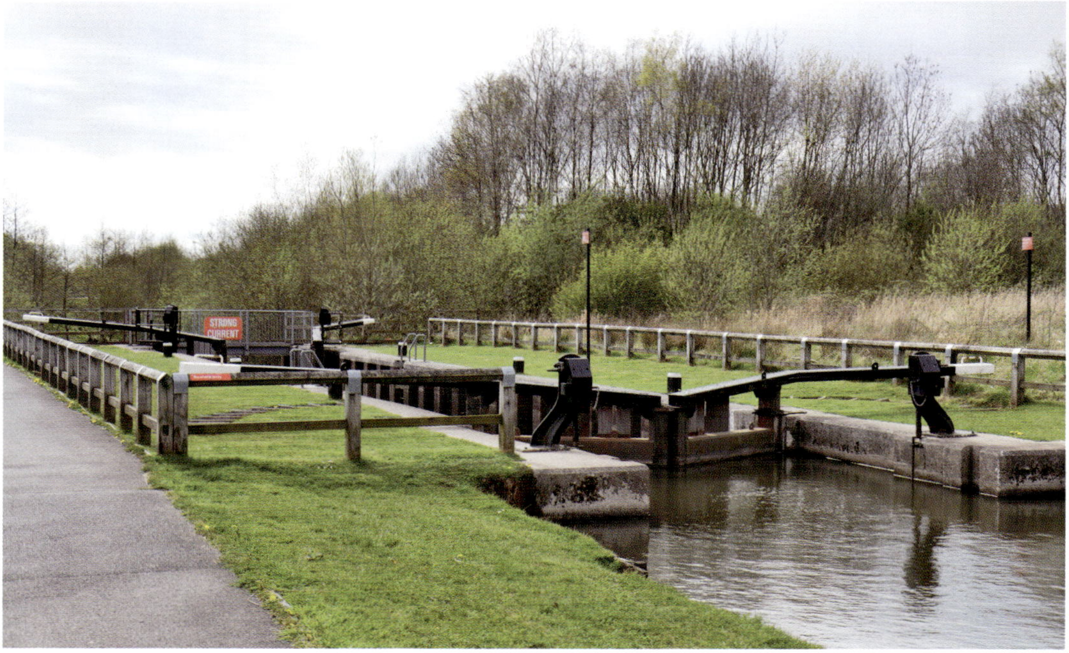

Millennium Ribble Link, Preston

The Ribble Link comprises eight locks that use, in part, the course of the Savik Brook from the Ribble to the Lancaster Canal. Pictured above is Lock four, and Lock one is shown below. Locks one to three comprise a triple staircase that brings craft up to the level of the Lancaster Canal. (*Ray Shill*)

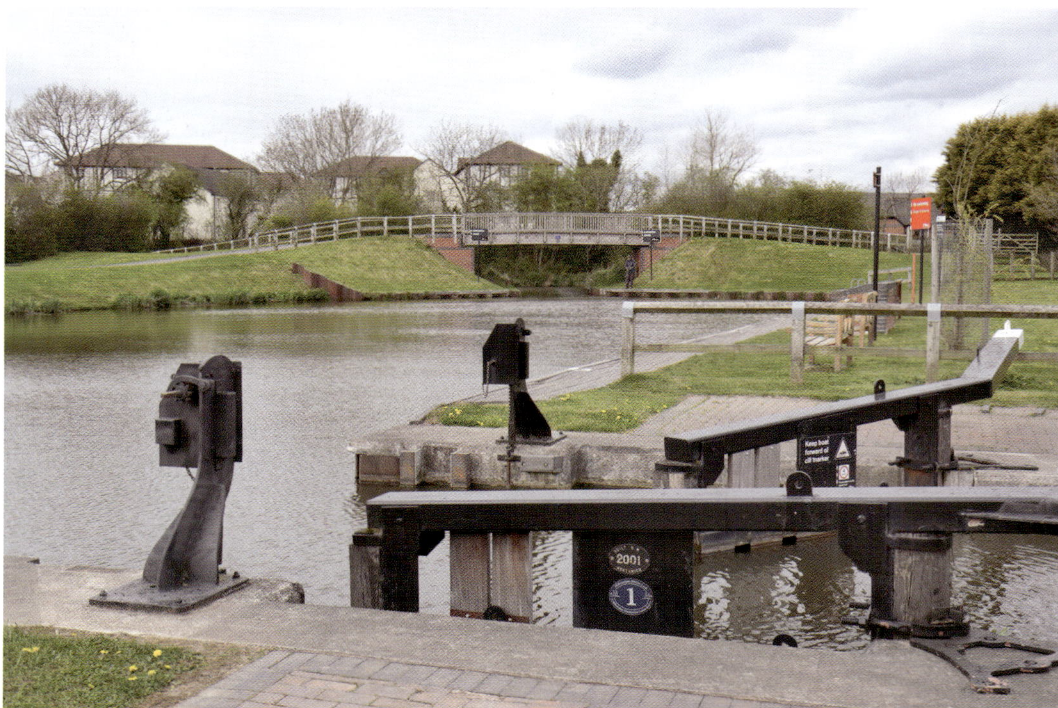

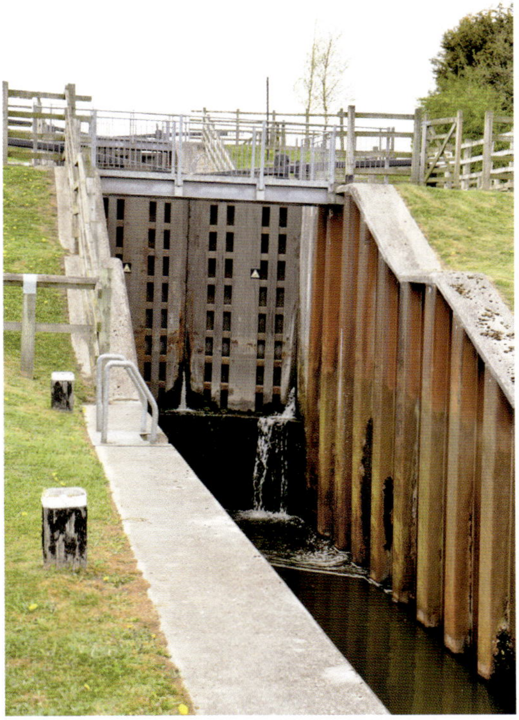

Millenium Ribble Link, Locks One to Three

Access to the Savik Brook Navigation required the making of a triple staircase lock in order to fit into the available plot of land. Boats travelling along this waterway have a tight turn at the bottom and boaters often prefer to reverse into the locks so that they can face forward for exit. (*Ray Shill*)

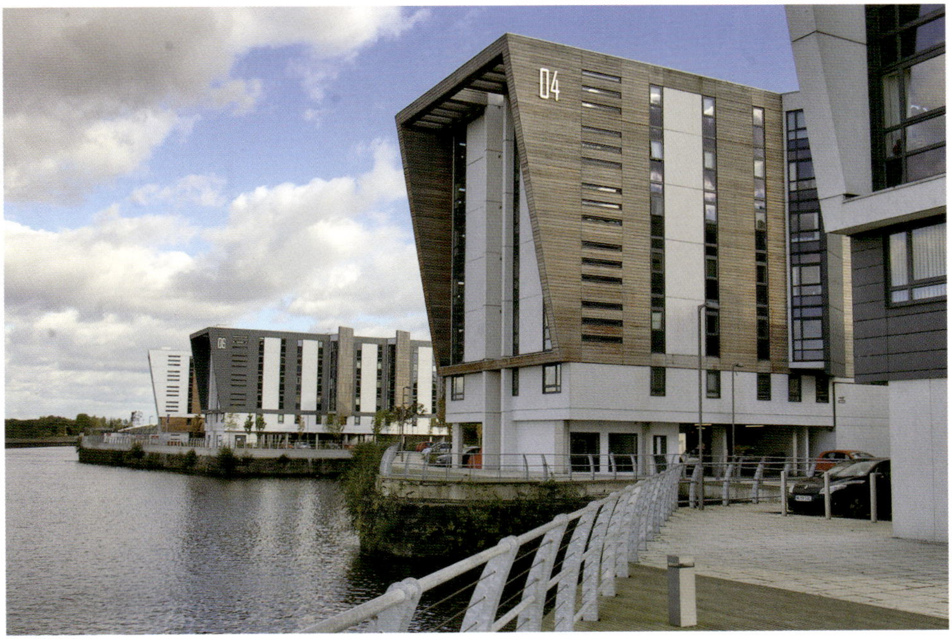

Old Quay, Runcorn, Manchester Ship Canal

While the making of the Manchester Ship Canal cut through the Latchford & Runcorn Canal at the Old Quay, part of the establishment was retained as canal workshops and some of the old lock walls also survived. The site has now been redeveloped for apartments that still retain the former waterfront walls. (*Ray Shill*)

Notable Structures on North West Canals

Grid references for lock flights and tunnels West to East, or North to South

Bridgewater Canal

Main line Stretford–Runcorn

Runcorn Docks	SJ499825-SJ503829
Runcorn (Old) Locks (W–E)	SJ04831-SJ510829
Runcorn (New) Locks (W–E)	SJ502828-SJ510829
Grand Junction Railway Aqueduct	SJ568811
Red Brow Aqueduct Underbridge	SJ569817
Sutton Heath Wharf	SJ613863
Waterloo Bridge	SJ510829

Preston Brook Branch

Preston Brook Wharf and Warehouses	SJ569804

Runcorn & Weston Canal

Parr's Lock	SJ497812
Railway Lock	SJ497818
Weston Point Junction	SJ497811

Dee

Chester Wharves	SJ397666-SJ397662
Saltney Wharves	SJ38664

Ditton Brook

Site of ironwork	SJ495809

Douglas Navigation

Douglas Chapel Lock

Gathurst Lock

Leeds & Liverpool Canal

Main Line

Litherland Swing Bridge	
Burscough Warehouses	SD443121
Douglas Aqueduct	
Appley Bridge Deep Lock	
Appley Locks	
Dean Locks	
Ell Meadow Lock	
Pagefield Lock	

Wigan Warehouses

Wigan Locks One to Twenty-Five

Rufford Branch

Latham Locks

Runnel Brow Lock

Moss Lock

German's Lock

Chicken Lock

Rufford Lock

Sollom (lock chamber)

Tarleton Lock

Stanley Dock Branch

Four locks (W–E)

Liverpool Link Lock Five

Liverpool Link Lock Six

Manchester Ship Canal

Eastham Locks

Ellesmere Port Wharf

Weston Marsh Lock

Weston Mersey Lock

Runcorn Docks

Old Quay Lock

Old Quay Swing Bridge

Moore Lane Swing Bridge

Acton Grange Viaduct

Chester Road Swing Bridge SJ604861

Norwich Road Swing Bridge SJ613864

Latchford High level bridge

Latchford Viaduct SJ633883

Latchford Locks

M6 Thelwall Viaduct SJ663883

Rixton Junction SJ682889

Mersey & Irwell Navigation

Rixton Junction to Warrington Bank Quay

Howley Wharf SJ620882

Warrington (Howley) Lock SJ616877

Latchford & Runcorn Canal

Twenty Steps Lock SJ613865

Wilderspool Turnbridge

Black Bear Wharf, Latchford Road Bridge

LNWR Bridge SJ618871

Manor Lock SJ625880

New Woolston Cut

Paddington Lock SJ628888

Woolston Lock SJ653888

Timber Feeder SJ629887

Old Woolston Cut

Old Woolston Cut Lock SJ648885

Walton Lock Branch, Manchester Ship Canal – Mersey & Irwell Navigation

Walton Lock SJ607864

Mersey

Bank Quay Wharf

Ditton Brook Wharves

Dukes Dock

Garston Dock

Runcorn Bridge

Salthouse Dock

Salisbury Dock – Access to Collingwood and Stanley Docks

Storeton Railway Pier

Sankey (St Helens) Canal

Widnes Lock	SJ515843
Swing Bridge	
Fiddler's Ferry Lock	
Sankey Lock	
Sankey Lift Bridge	SJ584876
Bewsey Lock Nine	
Hulme Lock Eight	
Winwick Wharf and Canal Workshops	
Winwick Lock Seven	
Hey Lock Six	
Bradley Lock Five	
Earlestown Newton Common Lock	
Engine Lock Three	
Old Double Lock Blackbrook	

Shropshire Union

Chester Canal

Barbridge Aqueduct	SJ614568
Barbridge Junction	SJ61259
Beeston Iron Lock	SJ556596
Beeston Stone Lock	SJ553598
Bunbury Staircase Lock	SJ577590
Calveley Interchange Wharf	SJ592587
Chemistry Lock	SJ435658
Chester & Crewe Railway Aqueduct	SJ433660
Christleton Lock	SJ420666
Greenfield Lock	SJ432651
River Gowy Aqueduct	
Hole Lane Lock	SJ417666
Nantwich Basin	SJ639627
Northgate Locks	SJ401666
Tarvin Lock	SJ426664
Tilstone Lock	SJ567593
Tower Wharf, Chester	SJ399666
Tower Wharf, Roving Bridge	SJ399667
Wharton's Lock	SJ538602

Dee Branch

River Lock	SJ398665

Bridge Lock SJ399666

Graving Lock SJ399667

Wardle Canal (Middlewich Branch)

Cholmondeston Lock SJ634576

Minshull Lock SJ654585

Minshull Aqueduct
(River Weaver) SJ664589

Stanthorne Lock SJ692657

Stanthorne Aqueducts (River
Wheelock and A530) SJ695657

Wirral Line (Ellesmere & Chester Canal)

Ellesmere Port SJ405775

Trent & Mersey Canal

Dutton Stop

Preston Brook Wharves

Barnton Tunnel

Preston Brook Tunnel

Saltisford Tunnel

Weaver Navigation

Acton Lock

Acton Swing Bridge

Acton Old Bridge

Butty Meadow Lock

Dutton Locks

Dutton Viaduct

Hartford Bridge

Hartford Lock

Hunt's Lock

Newbridge Lock

Northwich Swing Bridge

Northwich Lock

Northwich Viaduct

Pickerings Lock

Saltersford Lock

Witton Bridge Lock

Winnington Lock

Vale Royal Locks

Vale Royal Aqueduct

Anderton Lift Branch

Anderton Lift

Weston Point Branch

Railway Viaduct SJ534790

Weston Marsh Lock SJ505798

Weston Point Docks SJ495818-SJ495815

Waterman's Church SJ495815

Old Cut

Frodsham Lock SJ537782

Frodsham Bridge and
Wharf SJ530785

Frodsham Viaduct SJ530786

Weston Marsh Junction